IET ENERGY ENGINEERING SERIES 170

Reliability of Power Electronics Converters for Solar Photovoltaic Applications

Other volumes in this series:

Reliability of Power Electronics Converters for Solar Photovoltaic Applications

Edited by
Ahteshamul Haque, Frede Blaabjerg, Huai Wang,
Yongheng Yang, Zainul Addin Jaffery

The Institution of Engineering and Technology

Published by The Institution of Engineering and Technology, London, United Kingdom

The Institution of Engineering and Technology is registered as a Charity in England & Wales (no. 211014) and Scotland (no. SC038698).

The Institution of Engineering and Technology
Michael Faraday House
Six Hills Way, Stevenage
Herts, SG1 2AY, United Kingdom

www.theiet.org

British Library Cataloguing in Publication Data
A catalogue record for this product is available from the British Library

ISBN 978-1-83953-116-3 (hardback)
ISBN 978-1-83953-117-0 (PDF)

Typeset in India by Exeter Premedia Services Private Limited

Contents

6   **Control strategy for grid-connected solar inverter for IEC standards   141**
    Mohammed Ali Khan and Ariya Sangwongwanich

List of figures

List of tables

About the Editors

Frede Blaabjerg is a full professor at Aalborg University's Centre of Reliable Power Electronics, Denmark. His current research interests include power electronics and its applications such as in wind turbines, PV systems, reliability, harmonics, and adjustable speed drives. He has authored or co-authored more than 600 journal papers, and co-authored or edited fourteen books in power electronics. He is recipient of 32 IEEE Prize Paper Awards, the IEEE PELS Distinguished Service Award in 2009, the IEEE William E. Newell Power Electronics Award 2014, Global Energy Prize in 2019 and the 2020 IEEE Edison Medal. He has been President of the IEEE Power Electronics Society in 2019-2020. He was nominated in 2014-2020 by Thomson Reuters as one of the 250 most cited researchers in engineering in the world.

Ahteshamul Haque is an assistant professor at Jamia Millia Islamia University, New Delhi, India. His research focuses on power electronics and its application in renewable energy, drives, and other areas. Prior to Jamia Millia Islamia, he was working for a multinational organisation. He has received patents and awards for his work, and established an Advanced Power Electronics Research Lab.

Huai Wang is a full professor at the Centre of Reliable Power Electronics (CORPE), Aalborg University, Denmark. His research addresses the fundamental challenges in modeling power electronic component failure mechanisms and application issues in system-level predictability, condition monitoring, circuit architecture, and robustness design. He was previously a visiting scientist at the ETH Zurich, Switzerland, the Massachusetts Institute of Technology, USA, and with the ABB Corporate Research Centre. His awards include the 2016 Richard M. Bass Outstanding Young Power Electronics Engineer Award and the 2014 Green Talents Award from the German Federal Ministry of Education and Research.

Zainul Abdin Jaffery is a professor and Head of the Department of Electrical Engineering, Jamia Millia Islamia University, New Delhi, India. He has published about 80 research papers in the area of Electronics and Electrical engineering in journals and conferences. His research focuses on digital signal processing, digital image processing, and their applications in power engineering and electronics engineering. He is a senior member of IEEE (USA).

Yongheng Yang is a ZJU100 Professor at Zhejiang University, China. He received the Ph.D. degree from Aalborg University in 2014, where he was an associate professor in 2018-2020. He has published more than 250 scientific papers and two monographs. He received the 2018 IET Renewable Power Generation Premium Award. He was the IEEE Denmark Section Chair during 2019-2020. He is an associate editor for several IET/IEEE journals.

Chapter 1

Power electronics converters for solar PV applications

Ahteshamul Haque[1]

1.1 Introduction

The demand for renewable energy-based power plants is growing exponentially worldwide due to the climate change threat. The other reason for this demand is the exponential growth in electrical energy demand for industrialization. The availability of conventional fossil fuel reservoirs such as coal is very limited. Solar photovoltaic (PV)-based power plant is the most acceptable among all renewable energy sources as the sunlight is relatively available in abundance in most of the regions.

The total solar PV installed capacity is thus increasing worldwide, and it is forecasted that growth will continue, as shown in Figures 1.1 and 1.2, [2].

Almost in every business sector, the world is facing a growing recession, but in solar PV-based plant area, the job prospects have increased significantly, as indicated in Figure 1.3, and many opportunities are created at various levels, which helps in solving unemployment issues worldwide.

Based on the above data, it can be concluded that solar energy is the fastest growing renewable energy generation technique and most of the researchers are focusing on the development of a highly efficient and lossless method to achieve maximum power possible. Moreover, solar PV plants have technological challenges i.e. power conversion, quality factor, and safety issues. To address these issues various technological standards are made by regulating agencies, e.g., power electronics converters, control strategy, low-voltage ride-through (LVRT) [3, 4], anti-islanding [5, 6], and reliability, which will be discussed in the coming sections and chapters in detail.

[1]Advance Power Electronics Research Laboratory, Department of Electrical Engineering, Jamia Millia Islamia (A Central University), New Delhi, India

Solar PV Installed Capacity (Globally)

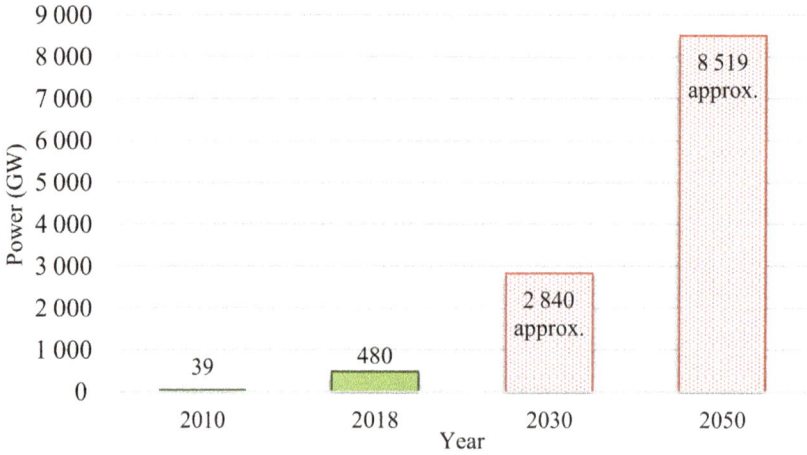

Figure 1.1   Solar PV installed capacity at world level [1]

1.2   Role of power electronics in solar PV systems

The layout of a solar PV power plant is shown in Figure 1.4. The solar PV output is DC and variable, i.e., it depends on ambient conditions (temperature, solar irradiance, etc.). These variable electrical signals need to be regulated as per the desired shape and magnitude. Power electronics converters are used to regulate the PV output. Moreover, power electronics converters are also used to convert the DC voltage into the AC voltage of the desired magnitude and frequency. The ability of

Solar PV Power Generation (% Power Generation Share–World Level)

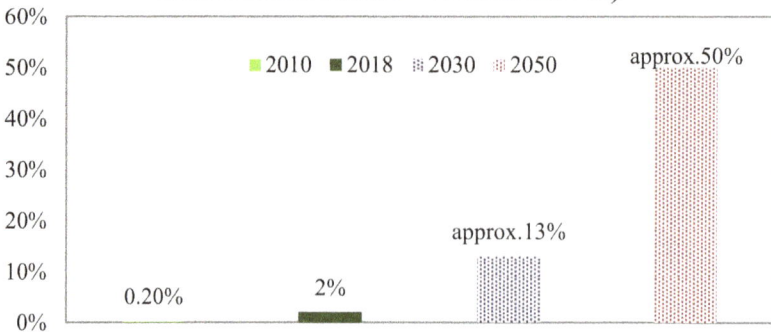

Figure 1.2   Power generation share of solar PV-based plants at world level [1]

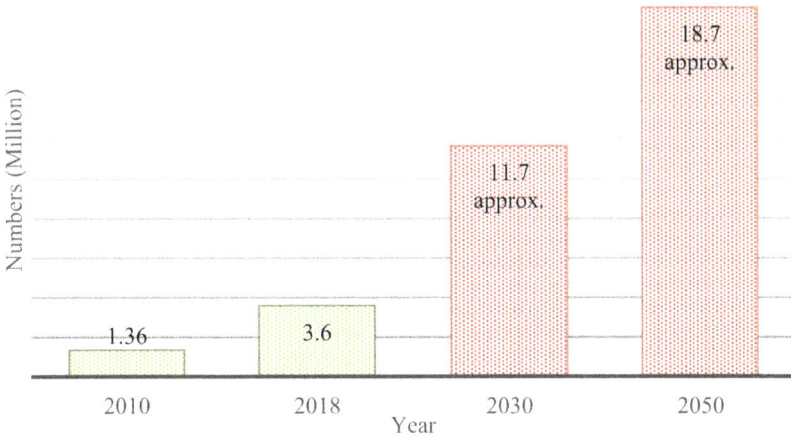

Figure 1.3   Employment from investment in solar PV plants [1]

power electronics switches to operate with pulse width modulation (PWM) makes the system-control efficient. Different control algorithms can be implemented for generating PWM and obtaining accurate output power with a low response time.

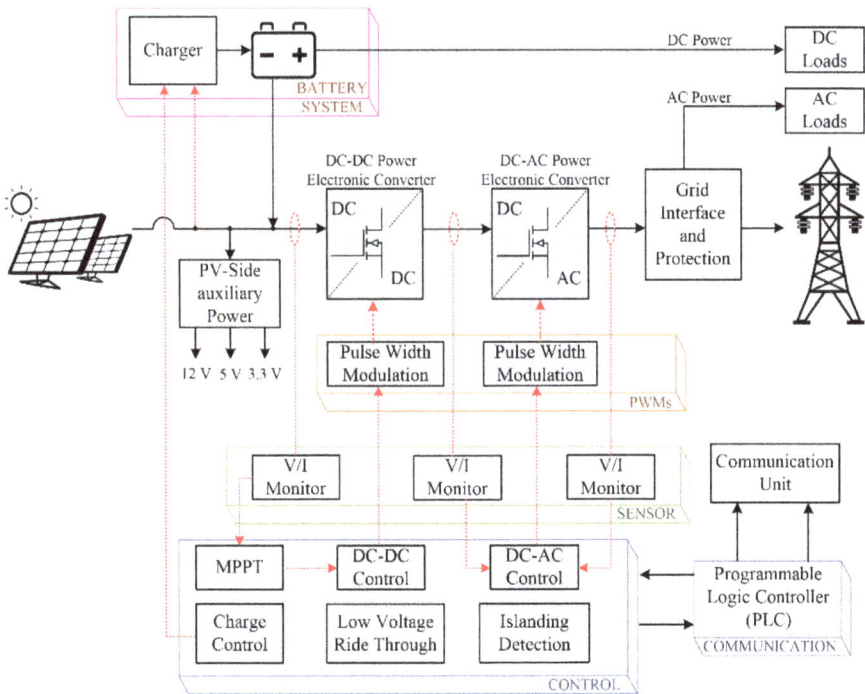

Figure 1.4   Layout of solar PV plant

Role of DC-DC Power Electronics Converters			
To regulate the DC output within limit	To implement MPPT and its control	To charge batteries	To derive the auxiliary power

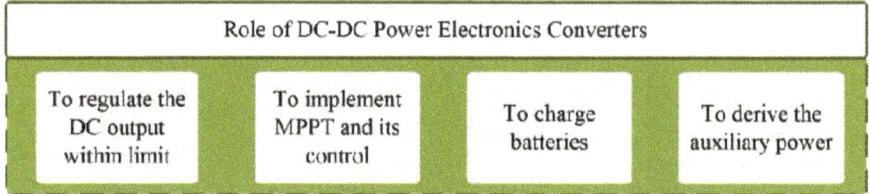

Figure 1.5   Main functions of DC–DC converters in solar PV plants

For a single-stage PV system, the output from the solar panel is directly fed to the DC–AC converter that regulated the incoming DC and converts it into AC. The output of the inverter is then passed through the filter before it is fed to the grid and local load. The control signal is generated by the measured voltage of the inverter at point of common coupling.

In the case of a two-stage system, the PV output from the solar panel is fed to the DC–DC converter. The maximum power point tracking (MPPT) control is implemented in the control of DC–DC converters. Also, if a battery is used in the solar PV plant, a charger circuit is required, which is also a power electronics converter. The PV-side auxiliary power is tapped off different DC voltage levels as per the requirement by the load. The main role of DC–DC converters in solar PV plant is shown in Figure 1.5. Similarly, the main functions of the DC–AC power electronics converter are shown in Figure 1.6. The LVRT and islanding detection protection are implemented in the control of the DC–AC converter.

1.3   DC–DC power electronics converters

In solar PV applications, DC–DC converters are used, and they are classified into two types: isolated and non-isolated type, as shown in Figure 1.7. The major difference between the two types of converters is in cost and galvanic isolation. The isolated type DC–DC has galvanic isolation between the input and output because of a magnetically coupled element, and at the same time, the cost is high. Flyback, resonant, forward, push-pull, bridge DC–DC converters are examples of isolated converters. The non-isolated type converters are Ćuk, SEPIC, boost, buck-boost, etc.

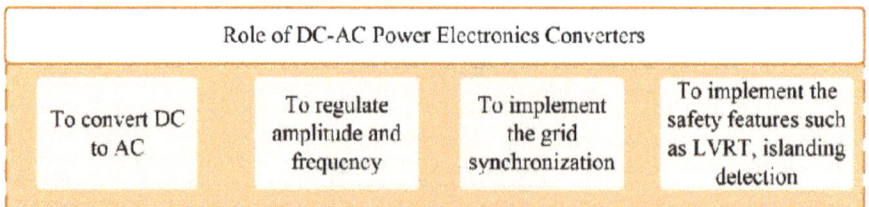

Role of DC-AC Power Electronics Converters			
To convert DC to AC	To regulate amplitude and frequency	To implement the grid synchronization	To implement the safety features such as LVRT, islanding detection

Figure 1.6   Main functions of DC–AC converters in solar PV plants

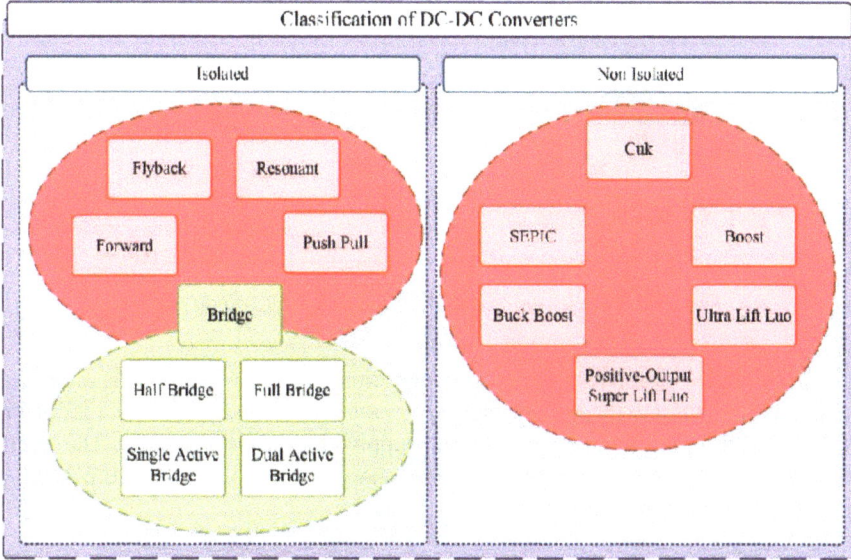

Figure 1.7   Classification of DC–DC converters used in solar PV applications

These DC–DC converters used in solar PV applications are discussed in Sections 1.3.1 to 1.3.14.

1.3.1   Buck converter

The buck converter, also known as a step-down converter, is used to step down the input voltage at the output side. The schematic of the converter is shown in Figure 1.8. This converter is a switched-mode power supply containing a diode, a power switch, and an energy storage element in the form of either a capacitor or a inductor. Here, the input voltage source feeds the controllable power switch

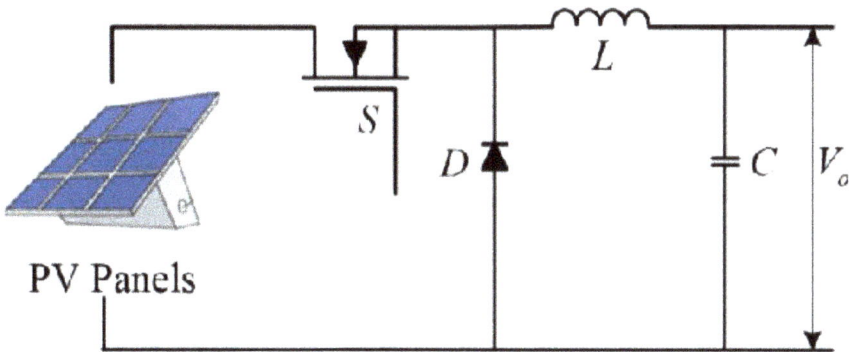

Figure 1.8   Schematic of a buck DC–DC converter

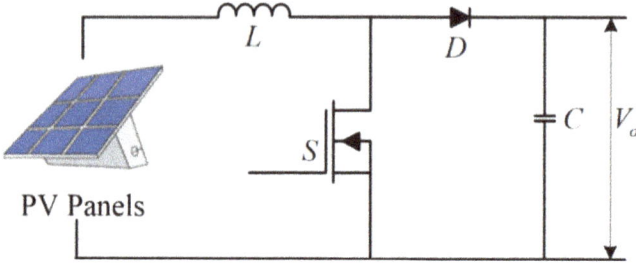

Figure 1.9   Schematic of the boost DC–DC converter

that is operated with a PWM either through a time base or with a frequency base. Generally, to eliminate the voltage ripple, a combination of capacitor and inductor can be used at both the load side and the supply side of the converter. The main application of this converter is in the battery charger circuit [7], solar PV pumping system [8], MPPT tracking, etc. [9].

1.3.2   Boost converter

The boost converter is referred to as a step-up converter and used in applications where the voltage magnitude at the output needs to be larger than the input voltage. This converter finds its majority of applications in PV systems to boost the PV voltage [10–12]. The schematic of the boost converter is shown in Figure 1.9. Similar to a buck converter, the boost converter is also a switched-mode power supply containing an inductor, a power switch, a diode, and a capacitor. Here, the input voltage source feeds the inductor that leads to a constant input current. Further, the power switch is operated with a PWM to achieve the required output voltage. Many modifications are available for the basic boost converter in the literature [10–13] to achieve ripple minimization, high voltage gain, and enhanced performance.

1.3.3   Buck-boost converter

The buck-boost converter can operate both in step-down and in step-up mode depending upon the duty cycle provided to the converter. The schematic of the converter shown in Figure 1.10 is developed by combining the basic buck converter

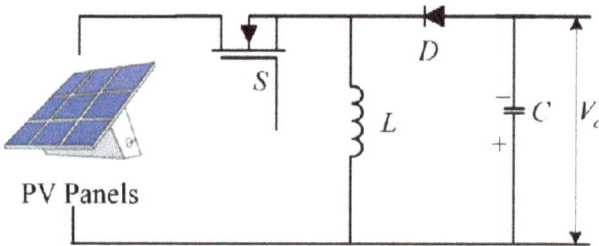

Figure 1.10   Schematic of the buck-boost DC–DC converter

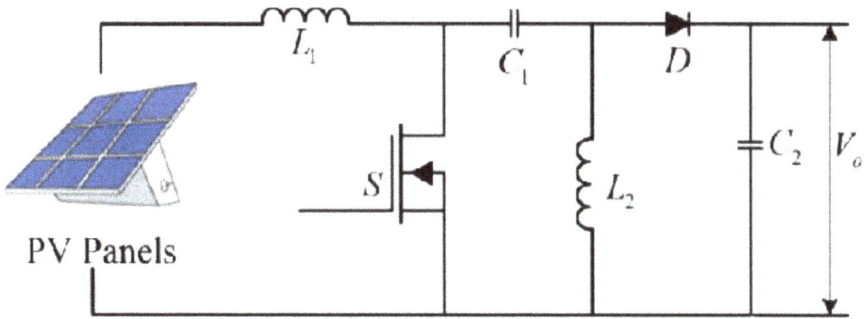

Figure 1.11   Schematic of the SEPIC DC–DC converter

and boost converter topologies discussed in Sections 1.3.1 and 1.3.2. This converter found most of its applications in stand-alone and grid-connected PV systems, and motor drives [14]. Similar to the operation of a buck converter, the input voltage source feeds the controllable power switch in the converter that is operated with a PWM. The literature identified that the continuous current mode operation of the buck-boost converter has lower ripples in the current. Further, a buck-boost converter with two power switches can have the least current and voltage stress on the components operating in the converter. Moreover, to enhance the operation of a basic buck-boost converter, various other topologies such as SEPIC [15], Ćuk [16], and Luo converters [17] are available in the literature.

1.3.4   Single-ended primary inductance converter

The single-ended primary inductance converter (SEPIC) is widely used in DC voltage flickering control and sensor-less control of PV applications. The schematic of the SEPIC is shown in Figure 1.11. While operating the converter, the on-time switching must be larger than the off-time switching to realize a high voltage at the output. Further, this also ensures that the capacitor is fully charged. For any condition, if the switching is not achieved, the converter fails to provide the required output. Besides, the operation of the converter along with a high-frequency transformer achieves an output voltage with minimized ripples. This provides various advantages with key features such as continuous output current, minimized output ripples, and minimized switching stress [15, 18–20].

1.3.5   Ćuk converter

The schematic of a Ćuk converter shown in Figure 1.12 has similarities with the basic buck-boost-converter except for the fact that the inductor is replaced with a capacitor to achieve the power transfer. This arrangement is also known as a negative-output capacitive energy-based flyback DC–DC converter [21]. Further, the Ćuk converter achieves ripple-free output in a system by inverting the output polarity of the converter with suitable connections [16, 22, 23]. Besides, to improve the efficiency of Ćuk converters and achieve optimal bidirectional operation concerning

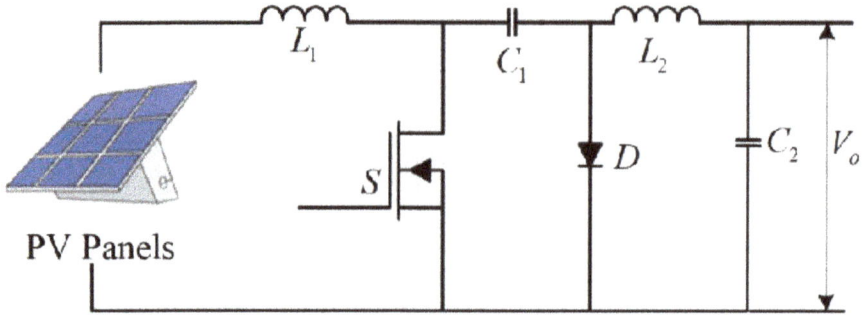

Figure 1.12   Schematic of a Ćuk DC–DC converter

the regulation of voltage and current [24], various modifications are proposed in the literature [25–28]. These studies have identified the application of the converter in various motor drive circuits [21] and renewable energy applications [29–31].

1.3.6   Positive-output super-lift Luo converter

The positive-output super-lift Luo converter shown in Figure 1.13 was initially introduced by Luo *et al.* [17] in 2003. This converter was developed with different series energy storage elements such as series inductors and capacitors that provide high output voltage resembling arithmetic progressions. Later, the design of the converter was modified in [32] by adding a high voltage transfer gain and its operation was enhanced by using a sliding mode controller in [33] for achieving the balance between the voltage regulation and load current. This converter is considered to be more powerful when compared to the SEPIC and Ćuk converters discussed in Sections 1.3.4 and 1.3.5 due to its unique features of enhanced efficiency and high

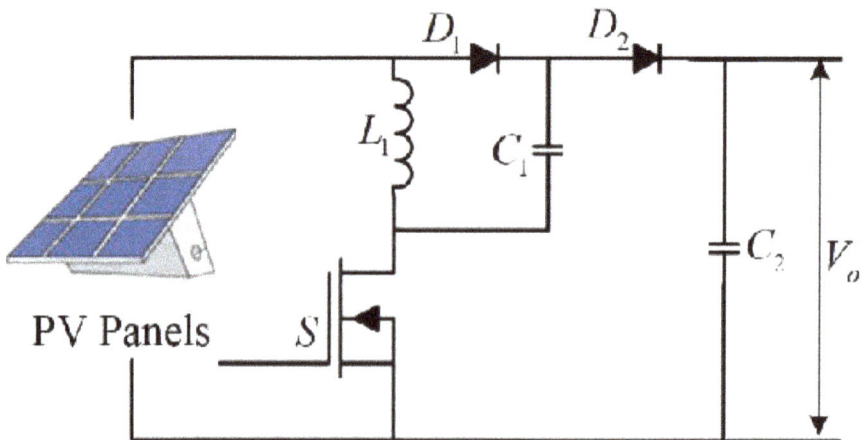

Figure 1.13   Schematic of a positive-output super-lift Luo DC–DC converter

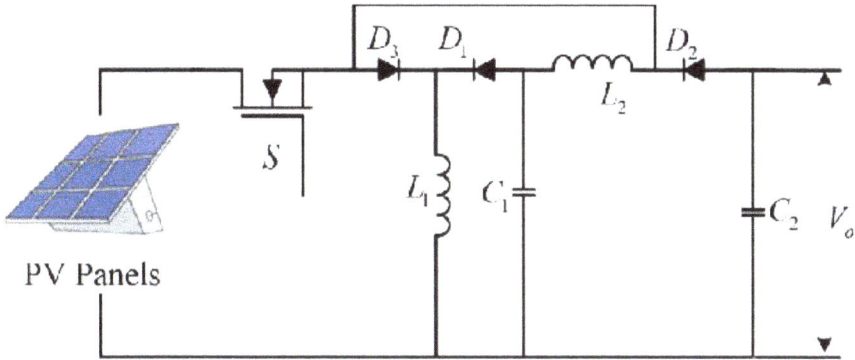

Figure 1.14   Schematic of an ultra-lift Duo DC–DC converter

output voltage resembling higher geometric progressions. Moreover, these converters are still under development for their operation with domestic and industrial PV
applications [34, 35].

1.3.7   Ultra-lift Luo converter

The ultra-lift Luo converter is shown in Figure 1.14 [36, 37]. This converter combines the design aspects of voltage and super-lift Luo converters to produce a high-
voltage conversion gain. This makes the converter highly efficient among the other
non-isolated DC–DC converters. Further, it is identified that the closed-loop design
of the converter is monotonous as the slightest variation in duty ratio results in large
output voltage variations.

1.3.8   Zeta converter

The Zeta converter combines the advantages of the buck-boost, SEPIC, and Ćuk
converters. The schematic of the Zeta converter is shown in Figure 1.15. When operated in a PV system, the Zeta converter enables continuous MPPT over the entire

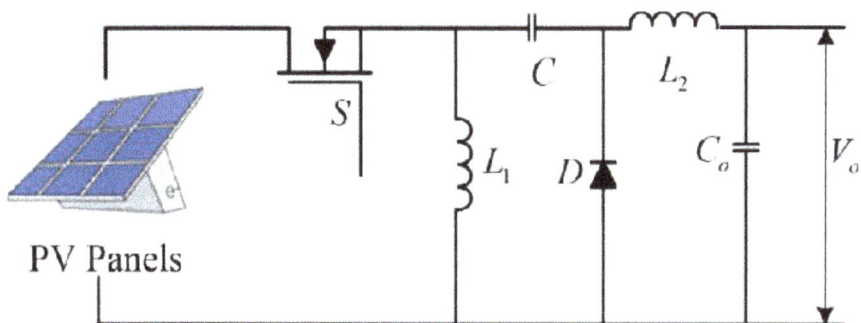

Figure 1.15   Schematic of a Zeta DC–DC converter

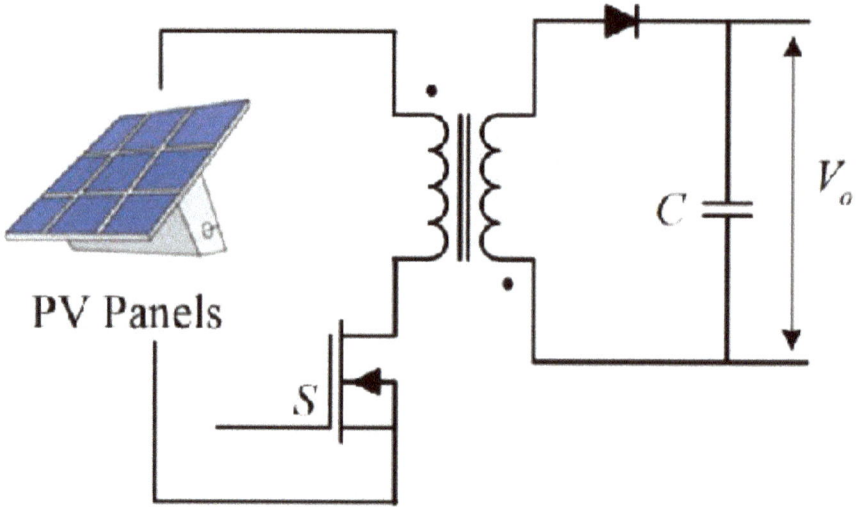

Figure 1.16   Schematic of a flyback DC–DC converter

area of the PV curve. Further, the Zeta converter provides a non-inverted output voltage that has either an enhanced or a diminished value concerning the input voltage [38–42]. Moreover, to reduce the output ripples and achieve enhanced voltage conversion in continuous and discontinuous modes, new topologies of the Zeta converter are developed in the literature [43]. These advancements are constituted for operation with the battery storage systems in PV applications.

1.3.9   Flyback converter

The schematic of the flyback DC–DC converter is shown in Figure 1.16. This converter acts as a key solution for higher converter gain requirements by employing transformers in the system. For a transformer with a large air gap to store energy, the flyback converter can be used in high-power applications. Further, the large air gap results in less magnetizing inductance, and the flyback converter provides very less energy transfer efficiency and large leakage flux. Moreover, the flyback converter overcomes the drawback of output polarity inversion and high current flow in the power switch and output diode in Ćuk converters [44]. These advantages have seen the application of flyback converters to operate in the discontinuous mode with isolated grid-connected inverters [45]. This application identified the unique features such as swift dynamic response and less complexity of the converter. Further, the efficiency and decreased ripple content of the converter is enhanced by employing various soft switching techniques [46, 47].

1.3.10   Three-port half-bridge DC–DC converter

The schematic of a three-port half-bridge DC–DC converter is shown in Figure 1.17. The converter's primary circuit operates in the buck converter mode

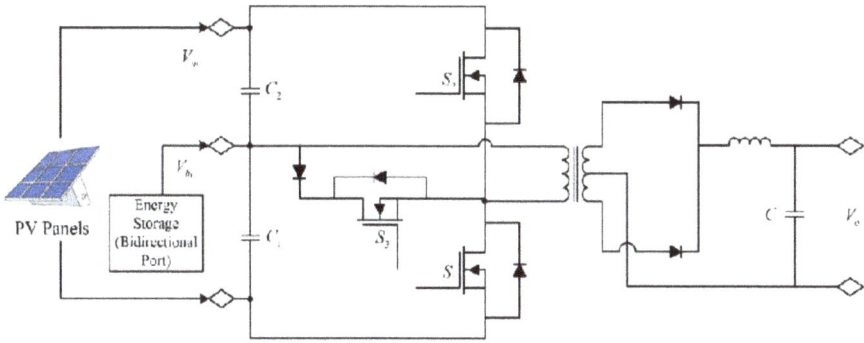

Figure 1.17   Schematic of a three-port half-bridge DC–DC converter

with synchronous rectification to provide the high-frequency transformer with a DC bias current. Further, various implementations are projected for post and synchronous regulations to regulate the three ports individually for achieving a single-stage power conversion with modest topology and simple control [48]. Moreover, to achieve continuous input current with a wide range of zero-voltage switching and low ripple, the three switches are operated with an active-clamped half-bridge DC converter [49]. Besides, to achieve a high voltage gain for large input voltage applications, the hybrid secondary rectifier is modified as a dual half-bridge LLC resonant converter. Here, depending on the switching strategy, the hybrid secondary rectifier acts as a quadruple rectifier [50]. In [51], the output power and efficiency are improved by employing an interleaved high-performance DC converter. A half-bridge of this topology ensures that the duty cycle of the two interleaved converters is close to 50 percent for achieving a continuous output current width with less lag and small component size. Further, to reduce the electromagnetic interference and voltage stress and achieve a high efficiency, the three-level converter is operated with a high-voltage bidirectional half-bridge in high-voltage DC microgrid applications [52].

1.3.11   Full-bridge converter

The full-bridge converter is used to integrate various components of the PV system such as PV array, energy storage device, and the load. The general schematic of a full-bridge converter is shown in Figure 1.18. The arrangement of a full-bridge converter consists of the integration of two buck-boost converters. This is developed to achieve zero-voltage switching and single power conversion with the topology [53]. Further, the circulating current losses are minimized by achieving zero-voltage switching during turn-on of switches through operating the full-bridge topology with asymmetrical PWM. Moreover, the use of asymmetrical PWM with a full-bridge converter minimizes the stress on power switches and achieves higher efficiency [54]. Besides, the problem of reverse recovery in output diode is overcome by achieving zero-current switchings during the switch turnoff through combining the resonant part of the circuit with the blocking capacitor and leakage inductance.

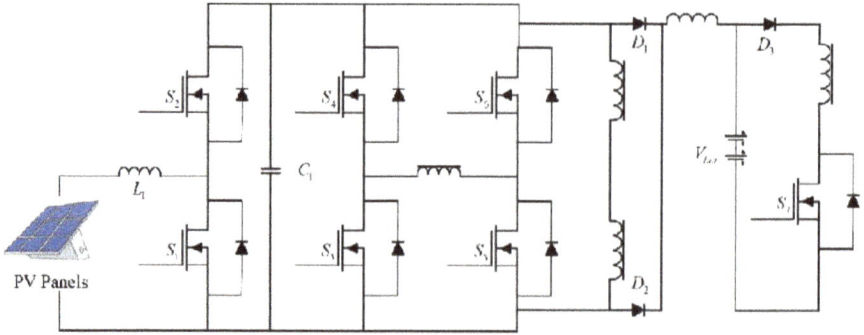

Figure 1.18   Schematic of a full-bridge DC–DC converter

1.3.12   Dual active bridge converter

The dual active bridge converter has found its application in stand-alone hybrid systems due to its advantages with high conversion efficiency, bidirectional power flow, galvanic isolation, and high power density [55–57]. The schematic of the converter is shown in Figure 1.19. From the circuit, it can be identified that the high-voltage DC sources feed the primary bridge and the low-voltage energy storage or load is connected to the secondary bridge. Further, a high-frequency power transformer is used to isolate the two full bridges whose leakage inductance is used as a storage element in the circuit. To enable the bidirectional power flow with the circuit, a square wave is conveniently phase shifted between both the bridges. Moreover, the voltage difference of the storage element is controlled to achieve power conversion with the circuit [58]. The control aspects of the dual active bridge-isolated bidirectional DC–DC converter are widely discussed through digital controllers in [59, 60]. In [61, 62], the high-frequency dual active bridge transformers were developed as an improvement to the existing converters. Besides, an ultra-capacitor-based dual active bridge converter was developed in [60].

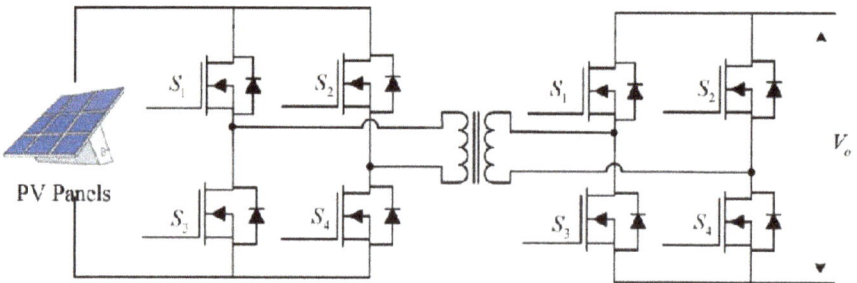

Figure 1.19   Schematic of a dual active bridge DC–DC converter

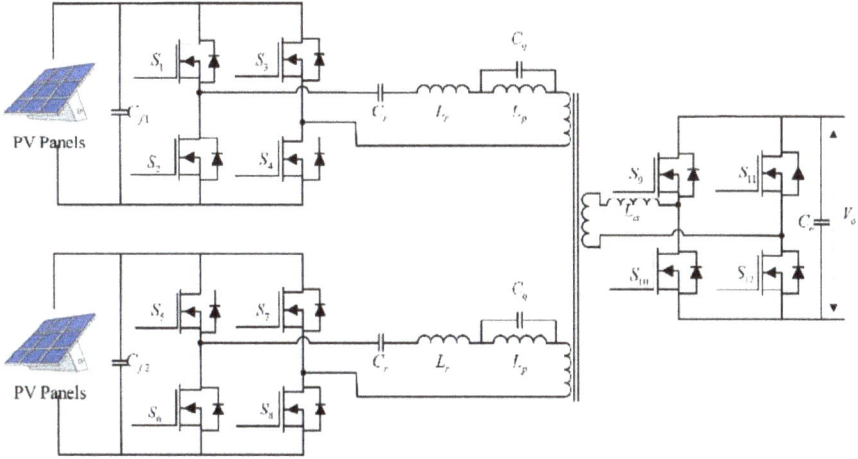

Figure 1.20   Schematic of a bidirectional multielement DC–DC resonant converter

1.3.13   Multielement resonant converter

The multielement resonant converter is an enhanced topology of the traditional LLC converter [63, 64]. This converter provides advantages of zero-voltage and -current switching using a short output circuit [65, 66], higher efficiencies at the full load operation, and high power density making them adaptable in renewable energy generation applications [67, 68]. The schematic of a three-port multielement resonant converter is shown in Figure 1.20. The design aspects of this converter involve series and parallel connection of five resonant components in the circuit. These multiple resonant components provide various resonant frequencies in the circuit and their suitable placement helps in transferring the active power of fundamental and third-order harmonics. Further, the parasitic leakage current due to the nonideal isolated transformer is ignored for this converter. From the literature, it is identified that the zero-voltage switching characteristics can be easily achieved for all the power switches in the three ports along with 96 percent power-conversion efficiency [69].

1.3.14   Push-pull converter

The push-pull converter, also known as a switching converter, is shown in Figure 1.21. This converter involves a transformer and operates with the help of a center-tapped primary winding by acting as a forward converter to the transformer core effectively. Further, the push-pull converter has small filters for different available power levels with the circuit. The major advantage of this converter is the transistor pair in the circuit employs input lines that avail the flow of current through the main winding of the transformer. Besides, the concurrent switching of the transistors draws current from the transformer resulting in a shattered condition for the current at the line during the switching condition of half-cycle pair. Moreover, when

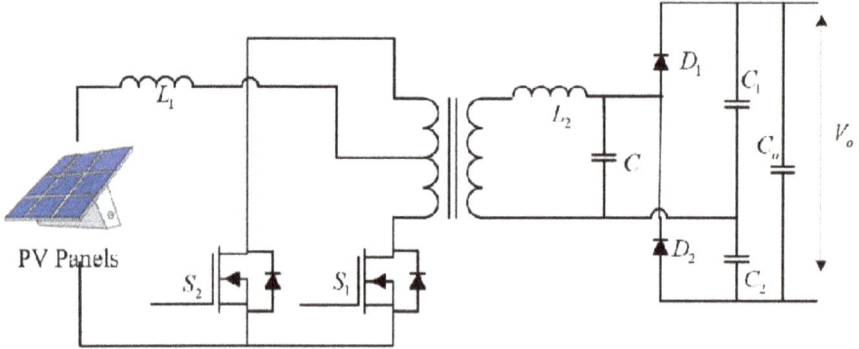

Figure 1.21   Schematic of a push-pull DC–DC converter

operated with low input noise, the push-pull converters have stable input current and can have efficient high-power applications [70]. Further, the center tapping that only utilizes half of the winding of the transformer at a time resulted in increased copper losses for the circuit.

Further, a summary of different DC–DC converters classified under isolated, non-isolated, unidirectional, and bidirectional are shown in Table 1.1 [71]. The summary identifies the important aspects of different converters and compares them with each other.

1.4   DC–AC converters

Solar inverters play a vital role in the PV application. The DC power from the panel after being boosted by a DC–DC converter is fed to the solar inverter before it can

Table 1.1   Comparison between different DC–DC converters

	Converter type	Optimal power demand	Voltage stress	Efficiency
Non-isolated unidirectional DC– DC converter	Buck	Low	High	Medium
	Boost	Low	High	Medium
	Buck-boost	Low	Medium	Medium
Non-isolated bidirectional DC– DC converter	Buck-boost	Low	Medium	Medium
	SEPIC	Low	High	Medium
	Ćuk	Low	High	Medium
	Half-bridge	Low	Medium	High
Isolated unidirectional DC–DC converter	Flyback	Low	High	High
	Half-bridge	Low	High	High
	Full-bridge	High	Medium	Medium
	Push-pull	High	High	Medium
Isolated bidirectional DC–DC converter	Half-bridge	Low	High	Medium
	Full-bridge	High	Low	Medium

be supplied to the load and grid. The inverter converts DC power into a regulated AC power that can be controlled by varying the switching of the power electronics devices. The basic inverter topologies are presented in Figure 1.22. The topologies are designed to operate in a grid-connected mode of operation while satisfying different grid codes requirements. From Figure 1.22, it can be deduced that the solar inverter can be categorized into two classes: transformer-based and transformerless inverters [72–74]. The galvanic isolation present in the transformer-based inverter provides isolation between the DC and the AC side and protects the system from any leakage power or power circulation that may occur in the AC side of the system. But the disadvantage of the transformer-based inverter is that the size of the inverter becomes large, and the efficiency of the inverter is low due to the transformer-based losses. To overcome the disadvantages, transformerless inverters were proposed. The absence of a transformer for the inverter reduces the size substantially and at the same time the efficiency of the inverter is improved [75]. However, the concern with transformerless inverters is the absence of isolation between the AC and DC side of the system. There is a change of power flow from AC to DC end, which may cause damage to DC equipment. As a result, many different topologies have been proposed to reduce leakage currents in transformerless inverters.

1.4.1   Transformer-based inverter

The transformer in the inverter provides galvanic isolation that separates the AC from the DC. The isolation avoids injection of DC current into the grid and stops leakage currents from the grid into the DC circuit. Transformer inverters can be categorized based on the operating frequency. The low-frequency transformer inverter as depicted in Figure 1.23a [76] is more reliable and cost-effective but at the same time, the efficiency is significantly poor. Whereas in the case of the high-frequency transformer inverter depicted in Figure 1.23b [77], the size of the inverter is small, and the inverter is more efficient compared to the low-frequency one. Because of the high-frequency operation, the components are constantly operating in high stress and the reliability is low [78, 79].

1.4.2   Transformerless inverter

The transformer present in the inverter achieves galvanic isolation and reduces the leakage currents from the grid. This helps in ensuring safety, but a substantial amount of power is lost because of the transformer, which reduces the efficiency of the inverter. Due to the large size and low efficiency of the transformer-based inverter, there has been a shift in research toward the transformerless inverters for PV applications. A few of the commonly used transformerless inverter topologies [80] are explained in this section. A comparative analysis of different transformerless inverters is presented in Table 1.2.

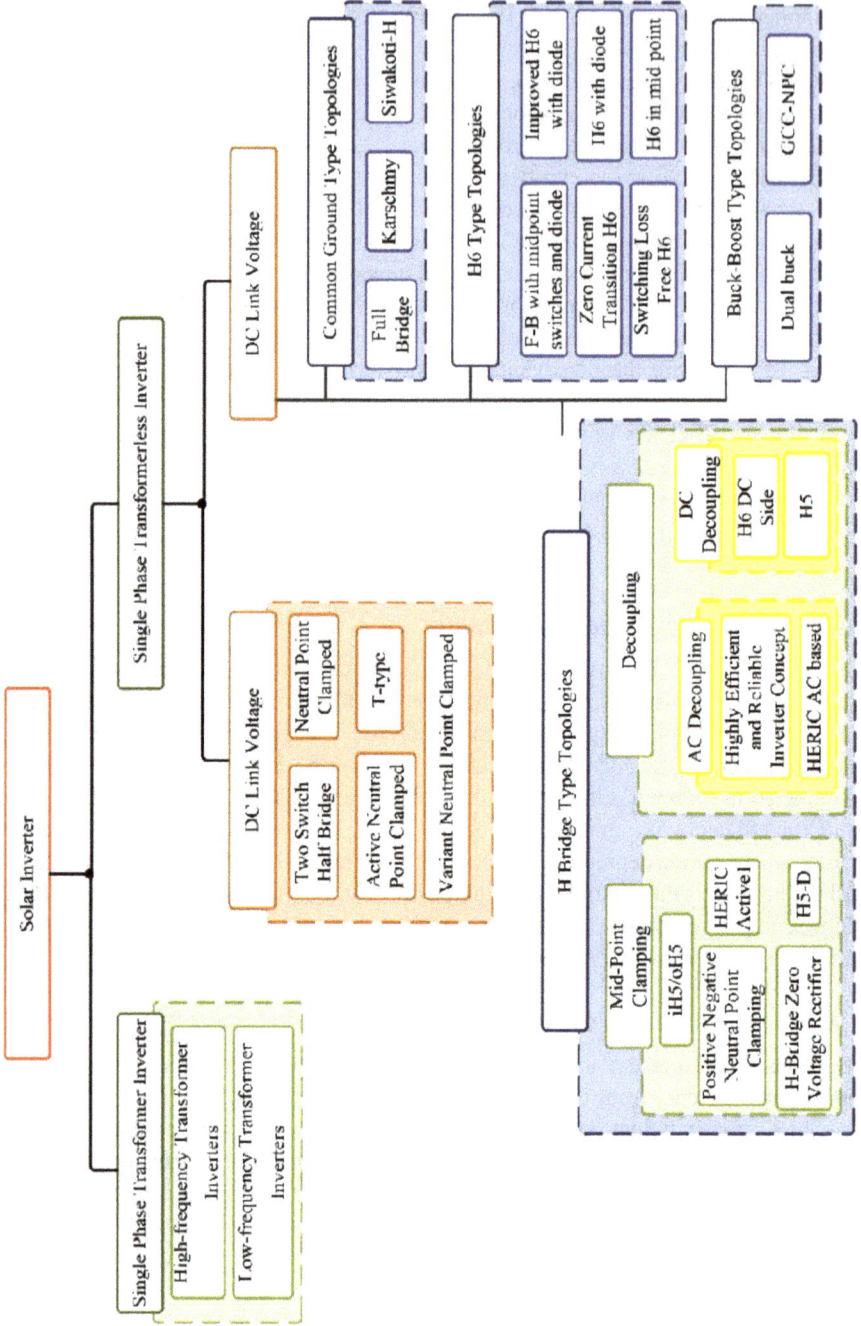

Figure 1.22   Classification of DC–AC converters used in solar PV application

(a)

(b)

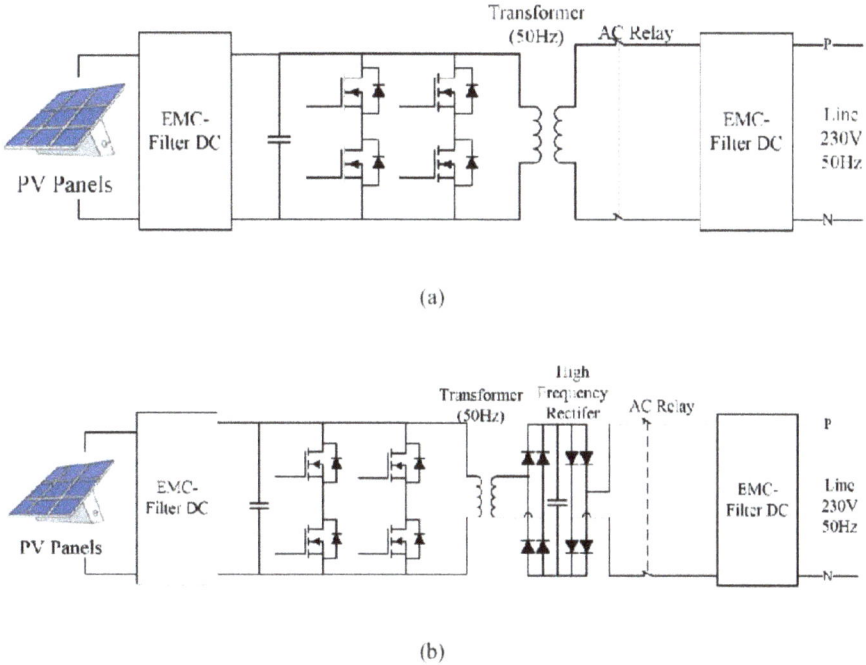

Figure 1.23   DC–AC converters with transformer: (a) low-frequency inverter, (b) high-frequency inverter

1.4.2.1   Half-bridge transformerless inverter

A half-bridge transformerless inverter comprises two power electronics switches connected in parallel with capacitors as shown in Figure 1.24 [105]. The operation is performed by turning on one switch at a time and charging the antiparallel capacitor at that instance. When the second switch is turned on, the corresponding capacitor is discharged, and the obtained inverter output is passed through an L-filter. The topology is simple to implement but achieving MPPT is difficult [106]. As a result, a high ripple is present in the output current.

1.4.2.2   Neutral point-clamped transformerless inverter

An neutral point-clamped (NPC) transformerless inverter comprises four power electronics switches with two diodes as shown in Figure 1.25 [82]. The clamping diodes at the midpoint aid in achieving the zero-voltage stage. Switches S_1 and S_3 operate in one-half cycle with alternating pulsing whereas S_2 and S_4 operate in another cycle. The current ripple is low compared to the half-bridge topology, but this topology is unable to balance conduction losses on the negative side causing a limitation for the DC-link [107].

Table 1.2 Comparison between different transformerless inverter topology

Single-phase transformerless inverter		Topologies	Components in topologies						Common-mode current	Common-mode voltage	Power factor (PF)	Total harmonic distortion (THD)	η (%)
			Semiconductor switches	Diodes	Passive element		Filter						
			Insulated-gate bipolar transistor		C	L	C	L					
DC-link voltage ($2V_{PV}$)		Two-switch H-B [81]	2	2	2	0	0	1	<20	Const.	N/A	1.1	N/A
		NPC [82]	4	2	2	0	0	1	=0	Const.	N/A	1.19	N/A
		ANPC [83]	6	6	2	0	0	1	<20	N/A	N/A	N/A	N/A
		T-type [84]	4	4	2	0	0	1	<20	Const.	N/A	N/A	N/A
DC-link voltage (V_{PV})	Common-ground-type topologies	S4 [85]	4	2	3	0	1	1	=0	Const.	0.8	2.1	97.2
		Karschmy [86]	5	2	2	1	1	1	=0	Const.	Unity	N/A	N/A
		Siwakoti-H [87]	4	1	2	0	1	1	=0	Const.	0.85	<2.3	97.8
		Improved H6 with diode [88]	6	2	1	0	1	2	<20	190–200	0.9	N/A	96.5
	H6-type topologies	ZCT-H6 [89]	8	3	4	2	1	2	<250	N/A	Unity	N/A	95.6
		H6 with diode [90]	6	2	1	0	1	2	<200	159–240	0.99	1.86	97.3
		SLF-H6 [91]	8	2	4	2	1	2	<150	N/A	Unity	N/A	96.2
		H6 in midpoint [92]	6	0	1	0	1	2	<200	159–240	Unity	N/A	N/A
	Buck-boost type topologies	Dual buck [93]	4	2	2	3	1	2	N/A	N/A	Unity	3.6	98.3
		Generation control circuit-neutral point clamped [94]	6	2	2	1	1	1	<20	N/A	Unity	4.08	95.7
	H-bridge type topologies — Midpoint clamping	iH5/oH5 [95]	6	0	2	0	1	2	<20	199–200	Unity	N/A	96.9
		H5-D [96]	5	1	2	0	1	2	<50	185–195	Unity	4.88	95
		HERIC Active 1 [97]	7	2	2	0	1	2	<25	199–200	N/A	N/A	N/A
		HB-ZVR [98]	5	5	2	0	1	2	<200	163–200	Unity	N/A	94.8
		PN-NPC [99]	8	0	2	0	1	2	<35	199–201	Unity	N/A	97.2
	Decoupling	HERIC [100]	6	0	1	0	1	2	<200	165–235	Unity	N/A	97.1
		HERIC AC-based [101]	6	2	1	0	1	2	<200	165–236	Unity	N/A	N/A
		H6 DC side [102]	6	0	2	0	1	2	<200	151–249	Unity	1.58	95.9
		H5 [103, 104]	5	0	1	0	1	2	<200	159–235	Unity	N/A	98.5

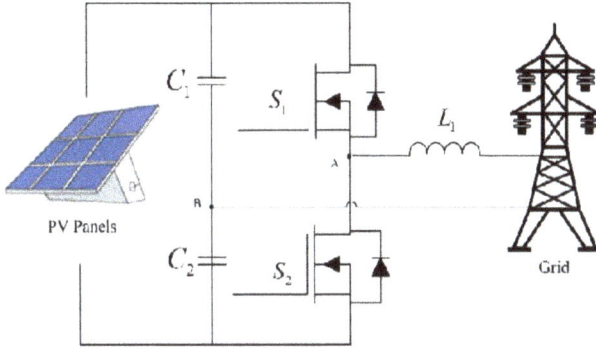

Figure 1.24   Two-switch half-bridge transformerless inverter

1.4.2.3   Active neutral point-clamped transformerless inverter

An active neutral point-clamped (ANPC) transformerless inverter illustrated in Figure 1.26 is a modification of an NPC inverter [108]. In the ANPC, the diodes are replaced by power electronics switches. The upper clamping is controlled by S_2 and S_5 whereas the lower clamping is regulated by S_3 and S_6 [109]. By replacing the diodes with the power electronics switches, the conduction losses of the inverter are controlled.

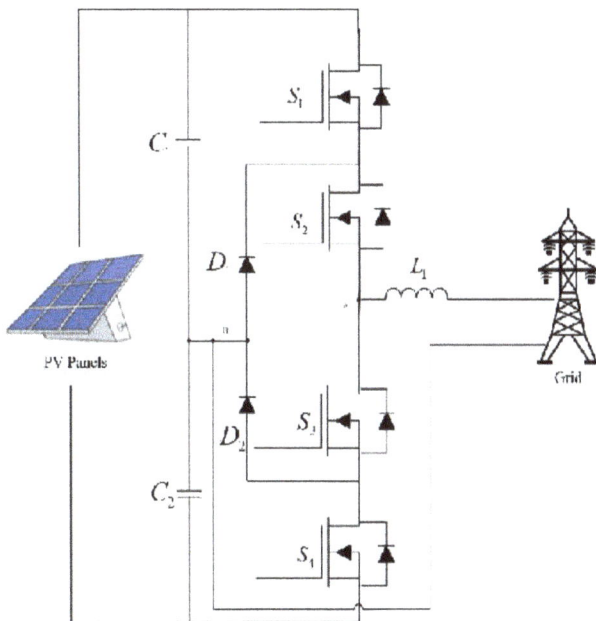

Figure 1.25   NPC transformerless inverter

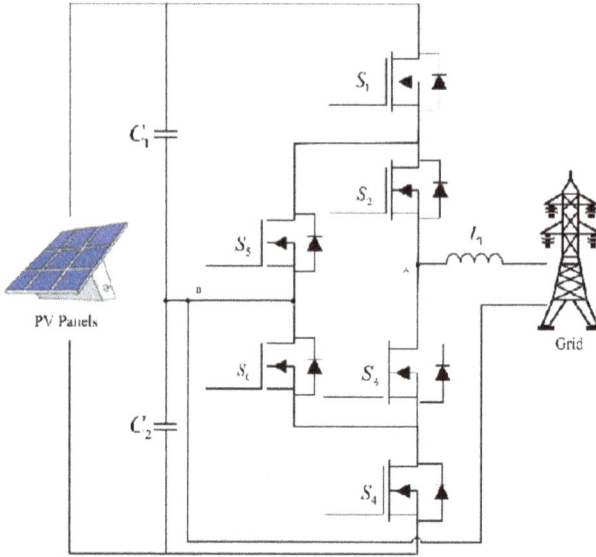

Figure 1.26   ANPC transformerless inverter

1.4.2.4   T-type transformerless inverter

A T-type three-level transformerless inverter consists of four power electronics switches with two bidirectional switches incorporated in the midpoint of the DC-link capacitor and switches (S_1 and S_2) leg [107] as illustrated in Figure 1.27. The S_1 and S_3 operate in a complementary way with S_2 and S_4 [110]. The clamping with two power electronics switches (S_3 and S_4) reduces the requirement of switching devices, which causes a reduction in conduction losses when compared to the ANPC topology.

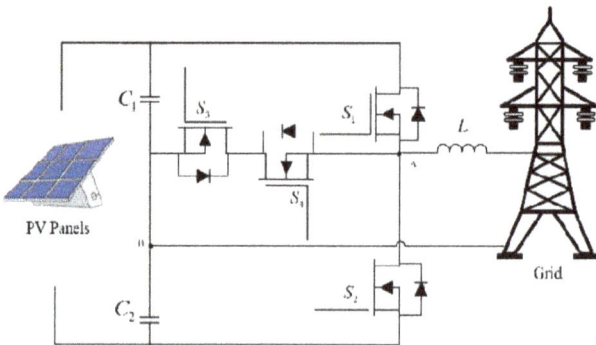

Figure 1.27   T-type transformerless inverter

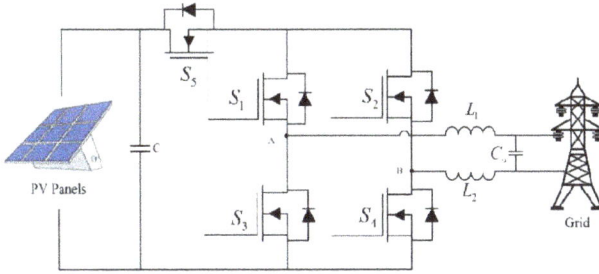

Figure 1.28   Three-switch transformerless inverter

1.4.2.5   Three-switch transformerless inverter

The three-switch transformerless inverter is a modified NPC or ANPC inverter. The number of power electronics switches is reduced from previous topologies, as shown in Figure 1.28. The topology incorporates a diode bridge along with S_3. The diode bridge provides the current path during the null state. The operation takes place in four modes. S_1 is turned on during the positive half-cycle whereas S_2 remains on during the negative half-cycle. During the freewheeling mode of a positive half-cycle, D_1 and D_4 are in forwarding bias with S_3 whereas for the negative half-cycle biasing, D_2 and D_3 are on alongside S_3 [111].

1.4.2.6   Full-bridge transformerless inverter

A full-bridge transformerless inverter consists of four power electronics switches as shown in Figure 1.29. During the positive half-cycle of operation, S_1 and S_4 are turned on and the antiparallel diode along S_2 and S_4 provide a path for the current flow. The S_2 and S_1 are complementary to each other and similarly S_3 and S_4 are complementary to each other. For positive half-cycle S_1 and S_4 are turned on and, as a result, the output voltage is equal to the input voltage. Whereas during the

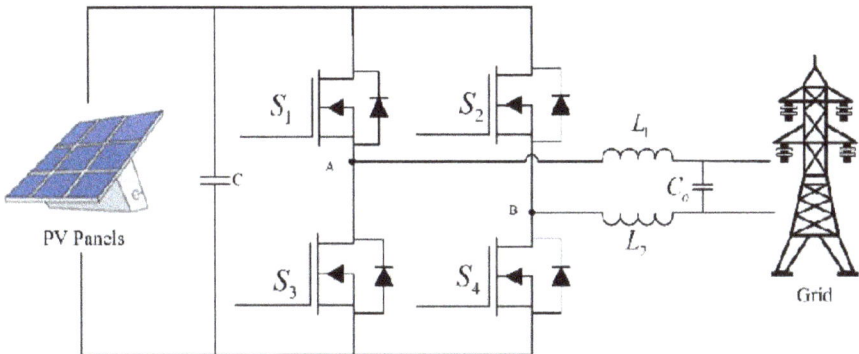

Figure 1.29   Full-bridge transformerless inverter

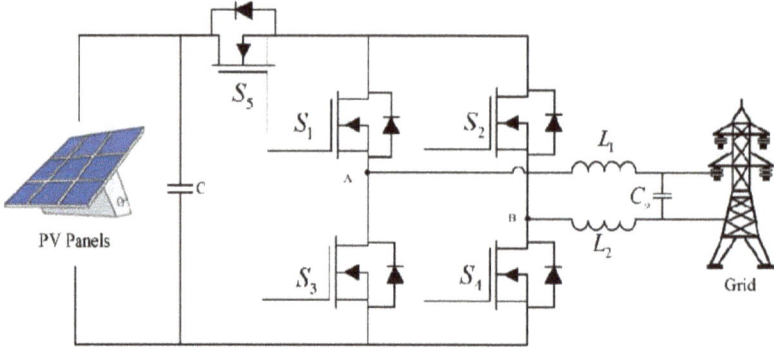

Figure 1.30   H5 transformerless inverter

freewheeling mode of the positive cycle, the current flows though S_1 and antiparallel diode of S_2 [112].

1.4.2.7   H5 transformerless inverter

The H5 transformerless inverter topology is the commonly implemented topology, which was patented and is commercially produced by SMA Solar Technology [103]. The topology consists of five power electronics switches as shown in Figure 1.30. It can be observed that S_5 on the DC side acts as a DC decoupling switch. During the freewheeling mode of operating, S_5 is turned off, which disconnects the DC side and effectively reduces the common-mode current [113, 114]. During the positive half-cycle, S_5 and S_4 operate at the switching frequency, whereas S_1 operates at the grid frequency. The other two switches remain in the off state. Whereas during the negative half-cycle, S_5 and S_2 operate at the switching frequency and S_3 operates at the grid frequency.

1.4.2.8   Full-bridge transformerless inverter with midpoint switches and diodes

The full-bridge transformerless inverter with midpoint switches and diodes topology is similar to the H5 transformerless inverter topology. Two extra switches are added on the top and bottom of a full-bridge inverter, as shown in Figure 1.31. While S_4 conduct, S_1 and S_6 conduct simultaneously. When S_3 is in on state, S_2 and S_5 operate with the same switching pulse. During a freewheeling period of a positive half-cycle, D_2 operates in forwarding bias along with S_4 . Whereas during the negative half-cycle, D_1 operates in forwarding bias along with S_2 [115].

1.4.2.9   H6-1 transformerless inverter

The H6-1 transformerless inverter consists of six power electronics switches, as shown in Figure 1.32 [116]. For operation in a positive half-cycle, S_1, S_4, and S_6 are turned on. And during the freewheeling state, S_6 along with the antiparallel diode of

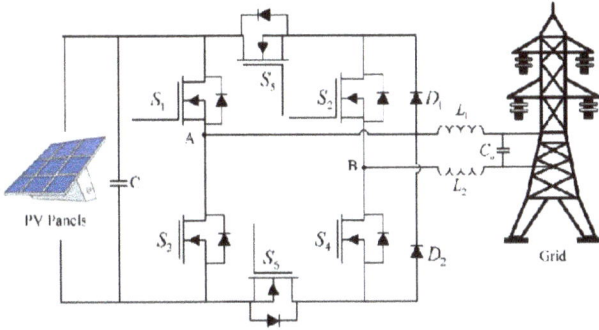

Figure 1.31   Full-bridge transformerless inverter with midpoint switches and diodes

S_5 conducts without any input, and the current flows through the load. During the negative half-cycle, S_2, S_3, and S_5 are turned on and for freewheeling state, S_5 along with the antiparallel diode of S_6 conducts.

1.4.2.10   Modified Highly Efficient and Reliable Inverter Concept

The HERIC transformerless inverter consists of six power electronics switches and has an issue with leakage currents. To overcome the drawback a modified version was proposed in [117–119] as shown in Figure 1.33. The topology aims at reducing the leakage current and keeping the common-mode voltage to the minimum. The drawback of this topology remains the shoot-through issue.

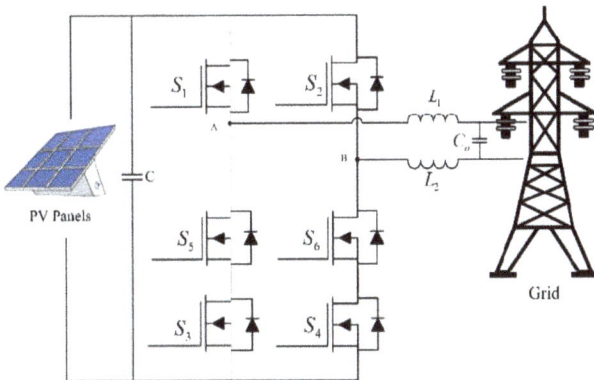

Figure 1.32   H6-1 transformerless inverter

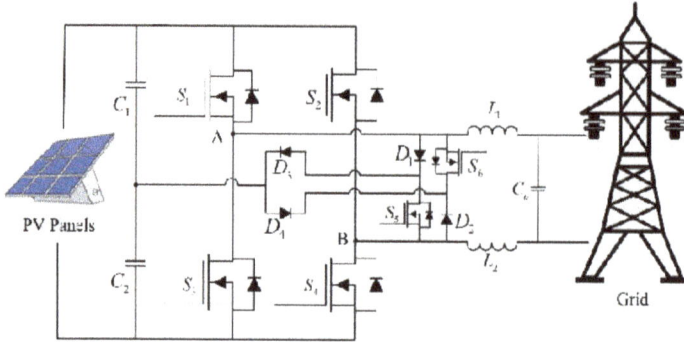

Figure 1.33   Modified HERIC transformerless inverter

1.4.2.11   oH5-1 transformerless inverter

The oH5-1 topology has been derived from the H5 transformerless inverter topology. The topology aims to clamp the input voltage to the half value [108]. The topology consists of six power electronics switches as shown in Figure 1.34.

1.4.2.12   Modified transformerless inverter

As illustrated in Figure 1.35, the topology presented is a modification over the basic full-bridge converter [120]. During the positive half-cycle S_1 and S_3 are in on state, whereas S_2 remains off. In case of the negative half-cycle, S_5 is in on state, whereas S_4 remains off.

1.5   Summary

Solar energy-based power generation is one of the major stakeholders in the renewable energy market around the globe. The low operating cost and easy availability

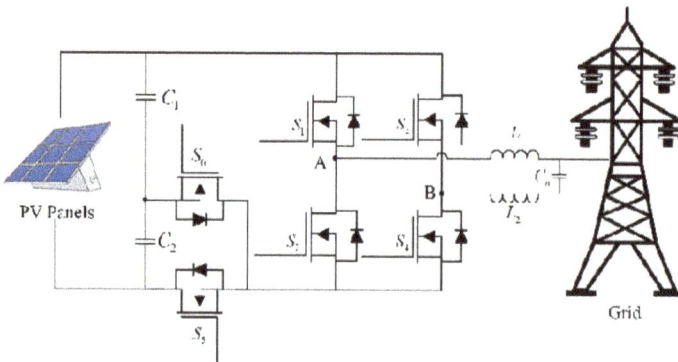

Figure 1.34   oH5-1 transformerless inverter

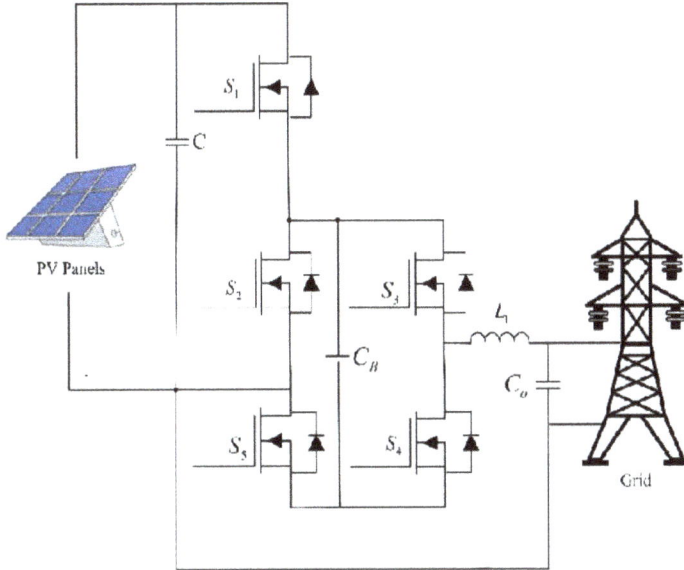

Figure 1.35   Modified transformerless inverter

have made it more favorable for the consumer. But many challenges are still present as a large amount of generated power is lost during the conversion process. As a result, many power electronics converters are designed over time to improve the efficiency and reliability of PV-based system. The chapter presents a brief introduction about the power electronics converters used for PV applications. Both the DC–DC converters and DC–AC converters are presented in this chapter and different topologies that are applied in PV applications are explained.

References

[1]   International Renewable Energy Agency – IRENA. *Future of solar photovoltaic: deployment, investment, technology, grid integration and socio-economic aspects*; 2019.
[2]   Zhao Y., Lehman B., De Palma J.F., Mosesian J., Lyons R. Challenges to overcurrent protection devices under line-line faults in solar photovoltaic arrays. 2011 IEEE Energy Conversion Congress and Exposition; Phoenix, AZ, USA, 17-22 Sept. 2011; 2011.
[3]   Fatama A.Z., Khan M.A., Kurukuru V.S.B., Haque A., Blaabjerg F. 'Coordinated reactive power strategy using static synchronous compensator for photovoltaic inverters'. *International Transactions on Electrical Energy Systems*. 2020;**30**(6):1–18.
[4]   Khan M.A., Haque A., Kurukuru V.S.B., Saad M. 'Advanced control strategy with voltage sag classification for single-phase grid-connected

photovoltaic system'. *IEEE Journal of Emerging and Selected Topics in Power Electronics*. 2020:1.

[5]   Khan M.A., Kurukuru V.S.B., Haque A., Mekhilef S. 'Islanding classification mechanism for grid-connected photovoltaic systems'. *IEEE Journal of Emerging and Selected Topics in Power Electronics*. 2020:1–1.

[6]   Fatama A., Haque A., Khan M.A. 'A multi feature based Islanding classification technique for distributed generation systems'. 2019 International Conference on Machine Learning, Big Data, Cloud and Parallel Computing; 2019. pp. 160–6.

[7]   Masoum M.A.S., Mousavi Badejani S.M., Fuchs E.F. 'Microprocessor-Controlled new class of optimal battery chargers for photovoltaic applications'. *IEEE Transactions on Energy Conversion*. 2004;**19**(3):599–606.

[8]   Elgendy M.A., Zahawi B., Atkinson D.J. 'Assessment of perturb and observe MPPT algorithm implementation techniques for PV pumping applications'. *IEEE Transactions on Sustainable Energy*. 2012;**3**(1):21–33.

[9]   Yusivar F., Farabi M.Y., Suryadiningrat R., Ananduta W.W., Syaifudin Y. 'Buck-converter photovoltaic simulator'. *International Journal of Power Electronics and Drive Systems*. 2011;**1**(2).

[10] Khazaei P., Mojtaba Modares S., Dabbaghjamanesh M., Almousa M., Moeini A. 'A high efficiency DC/DC boost converter for photovoltaic applications'. *International Journal of Soft Computing and Engineering*. 2016;**2**:2231–307.

[11] Saravanan S., Babu N.R. 'A modified high step-up non-isolated DC-DC converter for PV application'. *Journal of Applied Research and Technology*. 2017;**15**(3):242–9.

[12] Forouzesh M., Siwakoti Y.P., Gorji S.A., Blaabjerg F., Lehman B. 'Step-up DC–DC converters: a comprehensive review of voltage-boosting techniques, topologies, and applications'. *IEEE Transactions on Power Electronics*. 2017;**32**(12):9143–78.

[13] Huber L., Jovanovic M.M. 'A design approach for server power supplies for networking applications'. in APEC 2000. Fifteenth Annual IEEE Applied Power Electronics Conference and Exposition; 2000. pp. 1163–9.

[14] Howlader A.M., Urasaki N., Senjyu T., Yona A., Saber A.Y. 'Optimal PAM control for a buck boost DC–DC converter with a wide-speed-range of operation for a PMSM'. *Journal of Power Electronics*. 2010;**10**(5):477–84.

[15] Chiang S.J., Hsin-Jang Shieh., Ming-Chieh Chen. 'Modeling and control of PV charger system with SEPIC converter'. *IEEE Transactions on Industrial Electronics*. 2009;**56**(11):4344–53.

[16] Simonetti D.S.L., Sebastian J., dos Reis F.S., Uceda J. 'Design criteria for SEPIC and Cuk converters as power factor preregulators in discontinuous conduction mode'. Proceedings of the 1992 International Conference on Industrial Electronics, Control, Instrumentation, and Automation; San Diego, CA, USA, 13 Nov. 1992; 1992. pp. 283–8.

[17] Luo F.L., Ye H. 'Positive output super-lift converters'. *IEEE transactions on power electronics*. 2003;**18**(1):105–13.

[18] Bose B.K. 'Power electronics, smart grid, and renewable energy systems'. *Proceedings of the IEEE*. 2017;**105**(11):2011–18.

[19] Niculescu E., Niculescu M.C., Purcaru D.M. 'Modelling the PWM zeta converter in discontinuous conduction mode'. *MELECON 2008 – The 14th IEEE Mediterranean Electrotechnical Conference*; 2008. pp. 651–7.

[20] Al-Saffar M.A., Ismail E.H., Sabzali A.J., Fardoun A.A. 'An improved topology of SEPIC converter with reduced output voltage ripple'. *IEEE Transactions on Power Electronics*. 2008;**23**(5):2377–86.

[21] Song M.-S., Son Y.-D., Lee K.-H. 'Non-isolated bidirectional soft-switching SEPIC/ZETA converter with reduced ripple currents'. *Journal of Power Electronics*. 2014;**14**(4):649–60.

[22] Bist V., Singh B. 'PFC Cuk converter-fed BLDC motor drive'. *IEEE Transactions on Power Electronics*. 2015;**30**(2):871–87.

[23] Tse C.K., Lai Y.M., Iu H.H.C. 'HOPF bifurcation and chaos in a free-running current-controlled Cuk switching regulator'. *IEEE Transactions on Circuits and Systems I: Fundamental Theory and Applications*. 2000;**47**(4):448–57.

[24] 10.1109/TPEL.2016.2516255[Darwish A., Massoud A., Holliday D., Ahmed S., Williams B. 'Single-stage three-phase differential-mode buck-boost inverters with continuous input current for PV applications'. IEEE Transactions on Power Electronics; 2016. p. 1.

[25] Mostaan A., Baghramian A. 'Enhanced self lift zeta converter for negative-to-positive voltage conversion'. 4th Annual International Power Electronics, Drive Systems and Technologies Conference; 2013. pp. 212–17.

[26] Kamnarn U., Chunkag V. 'Analysis and design of a modular three-phase AC-to-DC converter using CUK rectifier module with nearly unity power factor and fast dynamic response'. *IEEE Transactions on Power Electronics*. 2009;**24**(8):2000–12.

[27] Lin B.-R., Chen J.-J., Shen S.-F. 'Zero voltage switching double-ended converter'. *IET Power Electronics*. 2010;**3**(2):187.

[28] Fardoun A.A., Ismail E.H., Sabzali A.J., Al-Saffar M.A. 'New efficient bridgeless cuk rectifiers for PFC applications'. *IEEE Transactions on Power Electronics*. 2012;**27**(7):3292–301.

[29] Durán E., Andújar J.M., Segura F., Barragán A.J. 'A high-flexibility DC load for fuel cell and solar arrays power sources based on DC–DC converters'. *Applied Energy*. 2011;**88**(no. 5):1690–702.

[30] Valencia P., Ramos-Paja C. 'Sliding-mode controller for maximum power point tracking in Grid-Connected photovoltaic systems'. *Energies*. 2015;**8**(11):12363–87.

[31] Jiménez-Castillo G., Muñoz-Rodríguez F.J., Rus-Casas C., Gómez-Vidal P. 'Improvements in performance analysis of photovoltaic systems: array power monitoring in pulse width modulation charge controllers'. *Sensors*. 2019;**19**(9):2150.

[32] Berkovich Y., Axelrod B., Madar R., Twina A. 'Improved Luo converter modifications with increasing voltage ratio'. *IET Power Electronics*. 2015;**8**(2):202–12.

[33] Kumar K.R., Jeevananthan S. 'Sliding mode control for current distribution control in paralleled positive output elementary super lift Luo converters'. *Journal of Power Electronics.* 2011;**11**(5):639–54.

[34] Luo F.L., Ye H. 'Positive output cascade boost converters'. *IEE Proceedings – Electric Power Applications.* 2004;**151**(5):590.

[35] Singh B., Bist V., Chandra A., Al-Haddad K. 'Power factor correction in bridgeless-Luo converter-fed BLDC motor drive'. *IEEE Transactions on Industry Applications.* 2015;**51**(2):1179–88.

[36] Luo F.L., Ye H. 'Ultra-lift Luo-converter'. *IEE Proceedings – Electric Power Applications.* 2005;**152**(1):27.

[37] Luo F.L., Ye H. 'Ultra-lift Luo-converter'. *2004 International Conference on Power System Technology, 2004 PowerCon.* **1**; 2004. pp. 81–6.

[38] Seyedmahmoudian M., Rahmani R., Mekhilef S., *et al.* 'Simulation and hardware implementation of new maximum power point tracking technique for partially shaded PV system using hybrid DEPSO method'. *IEEE Transactions on Sustainable Energy.* 2015;**6**(3):850–62.

[39] Narula S., Singh B., Bhuvaneswari G. 'Power factor corrected welding power supply using modified zeta converter'. *IEEE Journal of Emerging and Selected Topics in Power Electronics.* 2016;**4**(2):617–25.

[40] Singh B., Bist V. 'Reduced sensor configuration of brushless DC motor drive using a power factor correction-based modified-zeta converter'. *IET Power Electron.* 2014;**7**(9):2322–35.

[41] Singh S., Singh B., Bhuvaneswari G., Bist V. 'Power factor corrected zeta converter based improved power quality switched mode power supply'. *IEEE Transactions on Industrial Electronics.* 2015;**62**(9):5422–33.

[42] Murthy-Bellur D., Kazimierczuk M.K. 'Isolated two-transistor zeta converter with reduced transistor voltage stress'. *IEEE Transactions on Circuits and Systems II: Express Briefs.* 2011;**58**(1):41–5.

[43] A. M. S. S.A., Beltrame R.C., Schuch L., M. L. daS.M. 'PV module-integrated single-switch DC/DC converter for PV energy harvest with battery charge capability'. 2014 11th IEEE/IAS International Conference on Industry Applications; 2014. pp. 1–8.

[44] Gules R., dos Santos W.M., dos Reis F.A., Romaneli E.F.R., Badin A.A. 'A modified SEPIC converter with high static gain for renewable applications'. *IEEE Transactions on Power Electronics.* 2014;**29**(11):5860–71.

[45] D.D.-C.Lu., Agelidis V.G. 'Photovoltaic-battery-powered DC bus system for common portable electronic devices'. *IEEE Transactions on Power Electronics.* 2009;**24**(3):849–55.

[46] Achille E., Martire T., Glaize C., Joubert C. 'Optimized DC-AC boost converters for modular photovoltaic grid-connected generators'. *2004 IEEE International Symposium on Industrial Electronics.* **2**; 2004. pp. 1005–10.

[47] Hsieh Y.-C., Chen M.-R., Cheng H.-L. 'An interleaved flyback converter featured with zero-voltage transition'. *IEEE Transactions on Power Electronics.* 2011;**26**(1):79–84.

[48] Wu H., Chen R., Zhang J., Xing Y., Hu H., Ge H. 'A family of three-port half-bridge converters for a stand-alone renewable power system'. *IEEE Transactions on Power Electronics*. 2011;**26**(9):2697–706.

[49] Duong T.-D., Nguyen M.-K., Lim Y.-C., Choi J.-H. 'An active-clamped current-fed half-bridge DC-DC converter with three switches'. 2018 International Power Electronics Conference; 2018. pp. 982–6.

[50] Baek J.-I., Kim C.-E., Kim K.-W., Lee M.-S., Moon G.-W. Dual half-bridge LLC resonant converter with hybrid-secondary-rectifier (HSR) for Wide-Ouput-Voltage applications. 2018 International Power Electronics Conference (IPEC-Niigata 2018 – ECCE Asia); 2018. pp. 108–13.

[51] Hsieh H.-I., Chiu H.-L., Hsieh G.-C. 'Performance study of high-power half-bridge interleaved LLC converter'. 2018 International Power Electronics Conference; 2018. pp. 123–9.

[52] Andrijanovits A., Vinnikov D., Roasto I., Blinov A. 'Three-level half-bridge ZVS DC/DC converter for electrolyzer integration with renewable energy systems'. 2011 10th International Conference on Environment and Electrical Engineering; 2011. pp. 1–4.

[53] Hu W., Wu H., Xing Y., Sun K. 'A full-bridge three-port converter for renewable energy application'. *2014 IEEE Applied Power Electronics Conference and Exposition - APEC*. **2014**; 2014. pp. 57–62.

[54] Cha W.-J., Kwon J.-M., Kwon B.-H. 'Highly efficient asymmetrical PWM full-bridge converter for renewable energy sources'. *IEEE Transactions on Industrial Electronics*. 2016;**63**(5):2945–53.

[55] Stieneker M., De Doncker R.W. 'Dual-active bridge DC-DC converter systems for medium-voltage DC distribution grids'. 2015 IEEE 13th Brazilian Power Electronics Conference and 1st Southern Power Electronics Conference; 2015. pp. 1–6.

[56] Jeong D.-K., Kim H.-S., Baek J.-W., Kim J.-Y., Kim H.-J. 'Dual active bridge converter for energy storage system in DC microgrid'. 2016 IEEE Transportation Electrification Conference and Expo, Asia-Pacific; 2016. pp. 152–6.

[57] Sathishkumar P., Piao S., Khan M., *et al.* 'A blended SPS-ESPS control DAB-IBDC converter for a standalone solar power system'. *Energies*. 2017;**10**(9):1431.

[58] Ryu M., Jung D., Baek J., Kim H. 'An optimized design of bi-directional dual active bridge converter for low voltage battery charger'. 2014 16th International Power Electronics and Motion Control Conference and Exposition; 2014. pp. 177–83.

[59] Sathishkumar P., Krishna T., Himanshu., Khan M., Zeb K., Kim H.-J. 'Digital soft start implementation for minimizing start up transients in high power DAB-IBDC converter'. *Energies*. 2018;**11**(4):956.

[60] Haihua Zhou., Khambadkone A.M. 'Hybrid modulation for dual-active-bridge bidirectional converter with extended power range for ultracapacitor application'. *IEEE Transactions on Industry Applications*. 2009;**45**(4):1434–42.

[61] Zhao B., Song Q., Liu W., Sun Y. 'Overview of dual-active-bridge isolated bidirectional DC–DC converter for high-frequency-link power-conversion system'. *IEEE Transactions on Power Electronics*. 2014;**29**(8):4091–106.

[62] Naayagi R.T., Forsyth A.J., Shuttleworth R. 'High-power bidirectional DC–DC converter for aerospace applications'. *IEEE Transactions on Power Electronics*. 2012;**27**(11):4366–79.

[63] Liu R., Lee C.Q. 'Analysis and design of LLC-type series resonant convertor'. *Electronics letters*. 1988;**24**(24):1517.

[64] Koscelnik J., Frivaldsky M., Prazenica M., Mazgut R. 'A review of multi-elements resonant converters topologies'. *2014 ELEKTRO*; 2014. pp. 312–17.

[65] Sharifi S., Jabbari M., Farzanehfard H. 'A new family of single-switch ZVS resonant converters'. *IEEE Transactions on Industrial Electronics*. 2017;**64**(6):4539–48.

[66] Jabbari M., Kazemi H., Hematian N., Shahgholian G. 'A novel resonant LLC soft-switching buck converter'. *2014 IEEE 23rd International Symposium on Industrial Electronics*; 2014. pp. 370–4.

[67] Junjun Deng., Siqi Li., Sideng Hu., Mi C.C., Ruiqing Ma. 'Design methodology of LLC resonant converters for electric vehicle battery chargers'. *IEEE Transactions on Vehicular Technology*. 2014;**63**(4):1581–92.

[68] Khodabakhsh J., Moschopoulos G. 'A study of multilevel resonant DC-DC converters for conventional DC voltage bus applications'. 2018 IEEE Applied Power Electronics Conference and Exposition; 2018. pp. 2135–41.

[69] Wang Y., Han F., Yang L., Xu R., Liu R. 'A three-port bidirectional multi-element resonant converter with decoupled power flow management for hybrid energy storage systems'. *IEEE Access*. 2018;**6**:61331–41.

[70] Petit P., Aillerie M., Sawicki J.-P., Charles J.-P. 'Push-pull converter for high efficiency photovoltaic conversion'. *Energy Procedia*. 2012;**18**:1583–92.

[71] Turksoy O., Yilmaz U., Teke A. 'Overview of battery charger topologies in plug-in electric and hybrid electric vehicles'. *16th International Conference on Clean Energy*. 2018:9–11.

[72] Patrao I., Figueres E., González-Espín F., Garcerá G. 'Transformerless topologies for grid-connected single-phase photovoltaic inverters'. *Renewable and Sustainable Energy Reviews*. 2011;**15**(7):3423–31.

[73] Khan M.A., Haque A., Bharath K.V., Mekhilef S. 'Single phase Transformerless photovoltaic Inverter for grid connected systems – an overview'. *International Journal of Power Electronics*. 2018;**12**(1):1–28.

[74] Islam M., Mekhilef S., Hasan M. 'Single phase transformerless inverter topologies for grid-tied photovoltaic system: a review'. *Renewable and Sustainable Energy Reviews*. 2015;**45**(3):69–86.

[75] Gotekar P.S., Muley S.P., Kothari D.P., Umre B.S. Comparison of full bridge bipolar, H5, H6 and HERIC inverter for single phase photovoltaic systems - a review. 2015 Annual IEEE India Conference (INDICON); New Delhi, India, 17-20 Dec. 2015; 2015. pp. 1–6.

[76] Guo W., Jain P.K. 'A low frequency AC to high frequency AC inverter with build-in power factor correction and soft-switching'. *IEEE Transactions on Power Electronics*. 2004;**19**(2):430–42.

[77] Braun W.D., Perreault D.J. 'A high-frequency Inverter for variable-load operation'. *IEEE Journal of Emerging and Selected Topics in Power Electronics*. 2019;**7**(2):706–21.

[78] Kurukuru V.S.B., Haque A., Khan M.A., Tripathy A.K. 'Reliability analysis of silicon carbide power modules in voltage source converters'. 2019 International Conference on Power Electronics, Control and Automation; 2019. pp. 1–6.

[79] Shahzad M., Bharath K.V.S., Khan M.A., Haque A. 'Review on reliability of power electronic components in photovoltaic Inverters'. 2019 International Conference on Power Electronics, Control and Automation; 2019. pp. 1–6.

[80] Shayestegan M., Shakeri M., Abunima H., *et al.* 'An overview on prospects of new generation single-phase transformerless inverters for grid-connected photovoltaic (PV) systems'. *Renewable and Sustainable Energy Reviews*. 2018;**82**:515–30.

[81] Shimizu T., Hashimoto O., Kimura G. 'A novel high-performance utility-interactive photovoltaic inverter system'. *IEEE Transactions on Power Electronics*. 2003;**18**(2):704–11.

[82] Nabae A., Takahashi I., Akagi H. 'A new neutral-point-clamped PWM inverter'. *IEEE Transactions on Industry Applications*. 1981;**IA-17**(5):518–23.

[83] Yuan X., Stemmler H., Barbi I. Investigation on the clamping voltage self-balancing of the three-level capacitor clamping inverter. 30th Annual IEEE Power Electronics Specialists Conference. Record. (Cat. No.99CH36321); Charleston, SC, USA, 1 July 1999; 1999. pp. 1059–64.

[84] Zhang Z., Zhang J., Wu X. 'A single phase T-type inverter operating in boundary conduction mode'. 2016 IEEE Energy Conversion Congress and Exposition; 2016. pp. 1–6.

[85] Ardashir J.F., Sabahi M., Hosseini S.H., Blaabjerg F., Babaei E., Gharehpetian G.B. 'A single-phase transformerless inverter with charge pump circuit concept for grid-tied PV applications'. *IEEE Transactions on Industrial Electronics*. 2017;**64**(7):5403–15.

[86] Karschny D. 'Flying inductor topology, document de 196 42 522 C1'. 1998.

[87] Siwakoti Y.P., Blaabjerg F. 'H-bridge transformerless inverter with common ground for single-phase solar-photovoltaic system'. 2017 IEEE Applied Power Electronics Conference and Exposition; 2017. pp. 2610–4.

[88] Wang J., Luo F., Ji Z., *et al.* 'An improved hybrid modulation method for the single-phase H6 inverter with reactive power compensation'. *IEEE Transactions on Power Electronics*. 2018;**33**(9):7674–83.

[89] Xiao H.F., Lan K., Zhou B., Zhang L., Wu Z. 'A family of zero-current-transition transformerless photovoltaic grid-connected inverter'. *IEEE Transactions on Power Electronics*. 2015;**30**(6):3156–65.

[90] Ji B., Wang J., Zhao J. 'High-efficiency single-phase transformerless PV H6 inverter with hybrid modulation method'. *IEEE Transactions on Industrial Electronics*. 2013;**60**(5):2104–15.

[91] Xiao H.F., Zhang L., Li Y. 'An improved zero-current-switching single-phase transformerless PV H6 inverter with switching loss-free'. *IEEE Transactions on Industrial Electronics*. 2017;**64**(10):7896–905.

[92] San G., Qi H., Wu J., Guo X. 'A new three-level six-switch topology for transformerless photovoltaic systems'. Proceedings of The 7th International Power Electronics and Motion Control Conference; 2012. pp. *163–6*.

[93] Chen B., Sun P., Liu C., Chen C.-L., Lai J.-S., Yu W. 'High efficiency transformerless photovoltaic inverter with wide-range power factor capability'. 2012 Twenty-Seventh Annual IEEE Applied Power Electronics Conference and Exposition; 2012. pp. 575–82.

[94] Figueres E., Garcerá G., Patrao I., González-Medina R. 'Grid-tie inverter topology with maximum power extraction from two photovoltaic arrays'. *IET Renewable Power Generation*. 2014;**8**(6):638–48.

[95] Xiao H., Xie S. 'Leakage current analytical model and application in single-phase transformerless photovoltaic grid-connected inverter'. *IEEE Transactions on Electromagnetic Compatibility*. 2010;**52**(4):902–13.

[96] Li H., Zeng Y., Zhang B., Zheng T.Q., Hao R., Yang Z. 'An improved H5 topology with low common-mode current for transformerless PV grid- connected inverter'. *IEEE Transactions on Power Electronics*. 2018;**8993**.

[97] Li W., Gu Y., Luo H., Cui W., He X., Xia C. 'Topology review and derivation methodology of single-phase transformerless photovoltaic inverters for leakage current suppression'. *IEEE Transactions on Industrial Electronics*. 2015;**62**(7):4537–51.

[98] Kerekes T. *Analysis and Modeling of transformerless Photovoltaic Inverter*. Kjeller, Norway: Institut for Energiteknik, Aalborg Universitet; 2013.

[99] Xing Y. 'A family of neutral point clamped full-bridge topologies for transformerless photovoltaic grid-tied inverters'. *IEEE transactions on power electronics*. 2013;**28**(2):730–9.

[100] Araujo S.V., Zacharias P., Mallwitz R. 'Highly efficient single-phase transformerless inverters for grid-connected photovoltaic systems'. *IEEE Transactions on Industrial Electronics*. 2010;**57**(9):3118–28.

[101] Khan M.N.H., Forouzesh M., Siwakoti Y.P., Li L., Kerekes T., Blaabjerg F. 'Transformerless inverter topologies for single-phase photovoltaic systems: a comparative review'. *IEEE Journal of Emerging and Selected Topics in Power Electronics*. 2020;**8**(1):805–35.

[102] Yang B., Li W., Gu Y., Cui W., He X. 'Improved transformerless inverter with common-mode leakage current elimination for a photovoltaic grid-connected power system'. *IEEE Transactions on Power Electronics*. 2012;**27**(2):752–62.

[103] Victor M. 'US patent Application-H5 Inverter (SMA)'. 2005;**0286281**:A1.

[104] Khan M.A., Haque A., Kurukuru V.S.B. 'Voltage-balancing control for stand-alone H5 transformerless inverters'. Lecture Notes in Electrical Engineering; 2019. pp. 663–75.

[105] Kuo S.L. 'Half-bridge transistor inverter for DC power conversion'. *IEEE Transactions on Industrial Electronics and Control Instrumentation*. 1974;**IECI-21**(4):249–53.

[106] Gonzalez R., Gubia E., Lopez J., Marroyo L. 'Transformerless single-phase multilevel-based photovoltaic inverter'. *IEEE Transactions on Industrial Electronics*. 2008;**55**(7):2694–702.

[107] Schweizer M., Friedli T., Kolar J.W. 'Comparative evaluation of advanced three-phase three-level inverter/converter topologies against two-level systems'. *IEEE Transactions on Industrial Electronics*. 2013;**60**(12):5515–27.

[108] Tofigh Azary M., Sabahi M., Babaei E., Abbasi Aghdam Meinagh F. 'Modified single-phase single-stage grid-tied flying inductor Inverter with MPPT and suppressed leakage current'. *IEEE Transactions on Industrial Electronics*. 2018;**65**(1):221–31.

[109] Calais M., Agelidis V.G. 'Multilevel converters for single-phase grid connected photovoltaic systems – an overview'. *IEEE International Symposium on Industrial Electronics. Proceedings. ISIE'98*;1:224–9.

[110] Valderrama G.E., Guzman G.V., Pool-Mazun E.I., Martinez-Rodriguez P.R., Lopez-Sanchez M.J., Zuniga J.M.S. 'A single-phase asymmetrical T-type five-level transformerless PV inverter'. *IEEE Journal of Emerging and Selected Topics in Power Electronics*. 2018;**6**(1):140–50.

[111] Tang Y., Yao W., Loh P.C., Blaabjerg F. 'Highly reliable transformerless photovoltaic inverters with leakage current and pulsating power elimination'. *IEEE Transactions on Industrial Electronics*. 2016;**63**(2):1016–26.

[112] Wang J., Ji B., Zhao J., Yu J. 'From H4, H5 to H6 standardization of full-bridge single phase photovoltaic inverter topologies without ground leakage current issue'. *2012 IEEE Energy Convers Congr Expo ECCE 2012*. 2012:2419–25.

[113] Khan M.A., Haque A., Kurukuru V.S.B. 'Performance assessment of stand-alone transformerless inverters'. *International Transactions on Electrical Energy Systems*. 2019;**30**(1):1–21.

[114] Khan M.A., Haque A., Kurukuru V.S.B. 'Control and stability analysis of H5 transformerless inverter topology'. 2018 International Conference on Computing, Power and Communication Technologies; 2018. pp. 310–5.

[115] Kadam A., Shukla A. 'A multilevel transformerless inverter employing ground connection between PV negative terminal and grid neutral point'. *IEEE Transactions on Industrial Electronics*. 2017;**64**(11):8897–907.

[116] Mekhilef S., Islam M. 'H6-type transformerless single-phase inverter for grid-tied photovoltaic system'. *IET Power Electronics*. 2015;**8**(4):636–44.

[117] Khan A., Ben-Brahim L., Gastli A., Benammar M. 'Review and simulation of leakage current in transformerless microinverters for PV applications'. *Renewable and Sustainable Energy Reviews*. 2017;**74**(1):1240–56.

[118]   Khan M.A., Haque A., Kurukuru V.S.B. 'Intelligent control of a novel transformerless inverter topology for photovoltaic applications'. *Electrical Engineering*. 2020;**102**(2):627–41.

[119]   M. A. M. A.K., Haque A., Bharath K.V.S., Kurukuru V.S.B. 'Hybrid voltage control for stand alone transformerless inverter'. 2018 2nd IEEE International Conference on Power Electronics, Intelligent Control and Energy Systems; 2018. pp. 552–7.

[120]   Cavalcanti M.C., de Oliveira K.C., Neves F.A.S., Azevedo G.M.S., Camboim F.C. 'Modulation techniques to eliminate leakage currents in Transformerless three-phase photovoltaic systems'. *IEEE Transactions on Industrial Electronics*. 2010;**57**(4):1360–8.

Chapter 2

Wear-out failure prediction of a PV microinverter

Yanfeng Shen[1] and Huai Wang[2]

This chapter assesses and improves the reliability of a photovoltaic (PV) microinverter product by applying two different mission profiles and system-level electrothermal modeling. The system configuration and wear-out analysis process are described in brief before the electrothermal and lifetime models are developed for reliability-critical components in the microinverter, e.g., semiconductor devices and capacitors. Then the mission profiles of two distinct locations, Arizona, USA, and Aalborg, Denmark, are applied to the developed microinverter models, yielding the annual junction/hotspot temperature profiles and annually accumulative damages of components. Monte Carlo simulation and Weibull analysis are performed to obtain the system wear-out failure probability. Finally, an advanced multimode control scheme is introduced and a new long-lifetime electrolytic capacitor is employed in the DC link, leading to a significant reliability improvement of the PV microinverter product.

2.1 System description and reliability evaluation process

2.1.1 System description

The topology of the PV microinverter product is shown in Figure 2.1. As can be seen, it is composed of a synchronous quasi-Z-source series resonant DC–DC converter (qZSSRC) [1] and a grid-tied full-bridge inverter. Depending on the input voltage V_{pv}, three operation modes exist in the qZSSRC, i.e., pass-through mode (PTM), buck mode, and boost mode [1]. Thus, the qZSSRC can deal with a wide range of input voltages, e.g., from 10 V to 60 V. The developed 300-W product is shown in Figure 2.2, and its parameters are listed in Table 2.1. The full-load experimental waveforms and efficiencies of the PV microinverter prototype are shown in Figure 2.3. As can be seen, the microinverter operates well in all the three modes, and its efficiency can be maintained relatively high over wide input voltage and power ranges.

[1]Power Electronics, Department of Engineering – Electrical Engineering Division, University of Cambridge, United Kingdom
[2]Department of Energy Technology, Aalborg University, Denmark

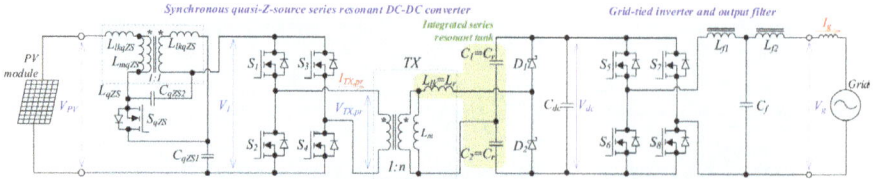

Figure 2.1   Schematic of the impedance-source PV microinverter product

2.1.2   Reliability evaluation process

It is identified in [2] that the temperature cycling is the most critical stress factor affecting the reliability of PV module-level power electronics. Therefore, this chapter takes into account the junction/hotspot temperature cycling and evaluates the wear-out failure probability of reliability-critical devices. The reliability evaluation flowchart is shown in Figure 2.4.

A real-field mission profile, including the ambient temperature and solar irradiance, is first translated into the microinverter input, i.e., the PV output voltage at maximum power points V_{mpp} and PV output power at maximum power points P_{mpp}. Then the power loss P_l and junction/hotspot temperature T_j can be obtained with an electrothermal model. The rainflow counting algorithm helps to derive the number of cycles of different stresses. With the component lifetime model and damage accumulation model, the accumulative damage can be obtained subsequently. In reality, there are many variations for component and model parameters, whose effect on the lifetime distribution can be evaluated with the Monte Carlo simulation. Finally, the system wear-out failure probability can be derived with the Weibull analysis and system reliability model.

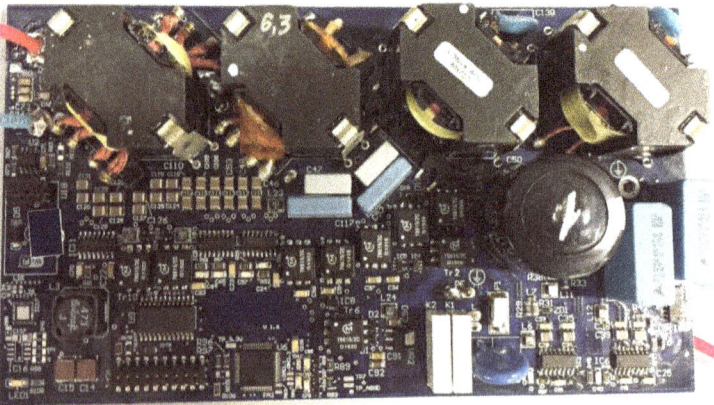

Figure 2.2   Photo of the PV microinverter product

Table 2.1   Specifications and parameters of the PV microinverter prototype

Descriptions	Parameters
Input voltage range	10–60 V
Nominal voltage	33 V
Most probable operating voltage	20–40 V
Rated power	300 W
Switching frequency of DC–DC stage	110 kHz
Switching frequency of inverter stage	20 kHz
Switches S_{qZS}, S_1–S_4	BSC035N10NS5
Switches S_5–S_8	SCT2120AFC
Diodes D_1–D_2	C3D02060E
Capacitors C_{qZS1} and C_{qZS2}	2.2μF × 12, C1210C225K1R
Coupled inductor L_{qZS}	L_{mqZS}= 12μH, L_{lkqZS}= 0.6μH, custom
Resonant capacitors C_1 and C_2	10 nF // 33 nF
	MKP1840310104M // B32672Z6333K
DC-link capacitor C_{dc}	150μF, 500-V electrolytic capacitor
Grid-side LCL filter: capacitor C_f	470 nF, B32653A6474K
Inductors L_{f1}	2.6 mH, custom
Inductors L_{f2}	1.8 mH, custom
Transformer TX	L_m= 1 mH, L_{lk}= 24μH, n= 6, custom

2.2   Electrothermal and lifetime modeling

2.2.1   Power loss modeling

The current stresses of critical components in the microinverter are measured under different PV output voltages and power levels, and thus the power
losses of MOSFETs, diodes, capacitors, and magnetic components can be calculated in different solar irradiances and ambient temperatures. It is noted
that the drain-source on-state resistance of a MOSFET, the voltage drop and
on-resistance of a diode, and the equivalent series resistance of an aluminum
electrolytic capacitor are temperature-dependent. Therefore, their junction/
hotspot temperatures interact with their power losses. The interdependency
is considered in the electrothermal modeling. For other components, e.g.,
ceramic and film capacitors (C_{qZS}, C_{r1}, C_{r2}, C_f) and magnetic components
(implemented with the 3C95 material), the dependency of their power losses
on the hotspot temperatures is not significant and is therefore neglected.

2.2.2   Thermal modeling

The microinverter prototype is enclosed in a naturally cooled aluminum case
with a dimension of 200 mm × 150 mm × 45 mm. The enclosure is filled
up with high-thermal-conductivity compound, and as a result, the thermal
cross-coupling (TCC) among the main devices is pronounced and cannot be
neglected. The thermal impedance network of an enclosed converter system is
shown in Figure 2.5. The thermal propagation from the junction to the ambient

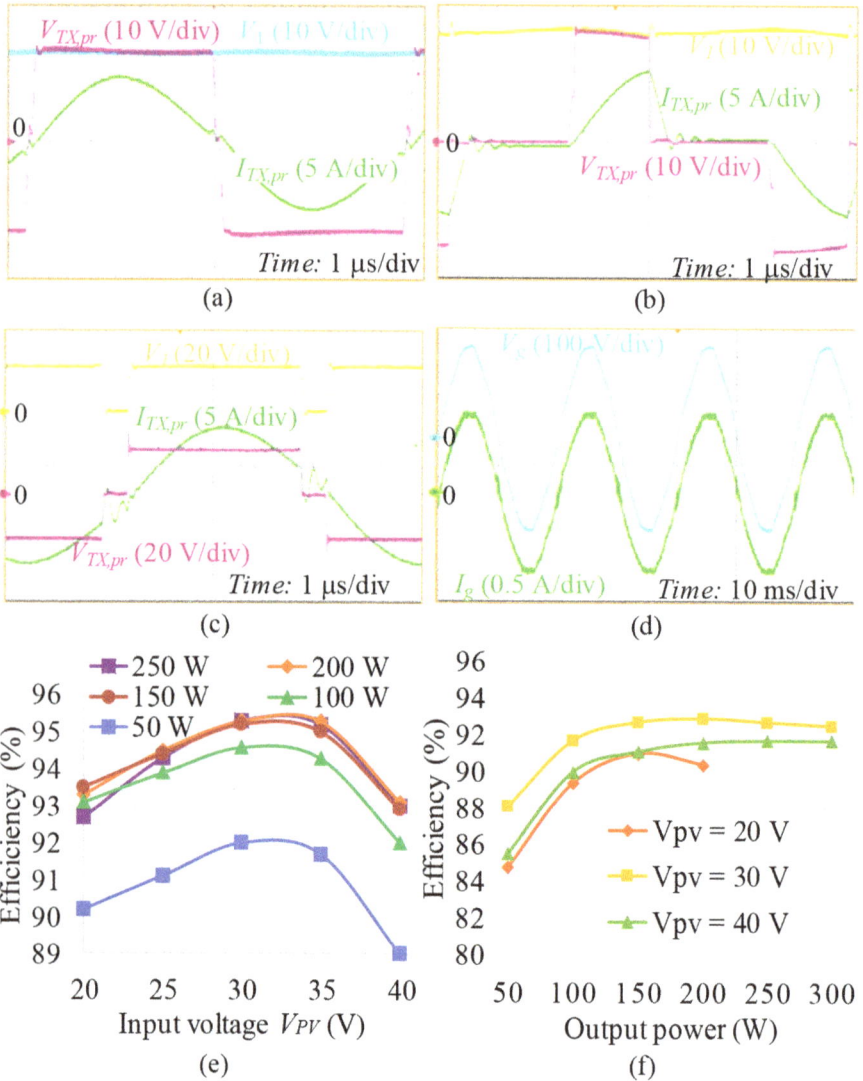

Figure 2.3   *Experimental waveforms in the (a) pass-through mode, (b) buck mode, and (c) boost mode. (d) Measured grid voltage and current waveforms. Measured efficiency curves of (e) the DC–DC stage and (f) the whole microinverter including the auxiliary power supply.*

can be divided into two parts, i.e., from the junction to the enclosure and from the enclosure to the ambient.

Considering the Printed Circuit Board (PCB) traces and real physical properties of components, a finite element method (FEM) structure model of

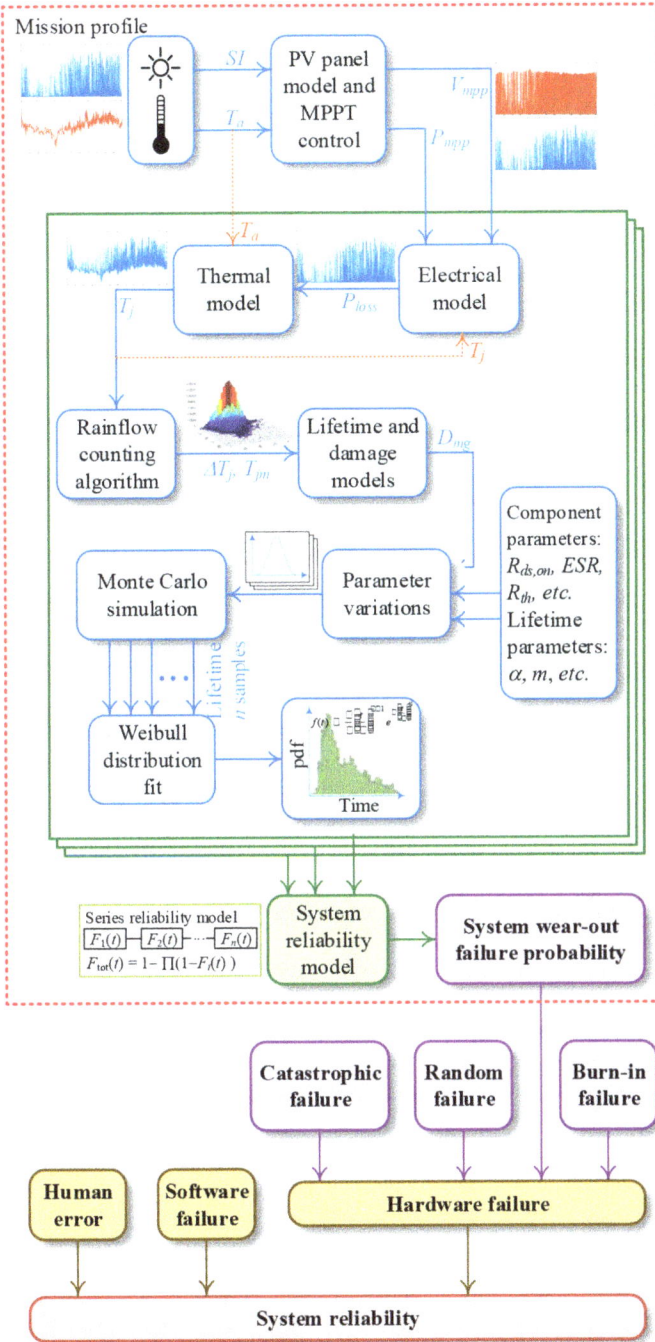

Figure 2.4   Failure modes of power electronics systems and evaluation flowchart of the hardware wear-out failure probability

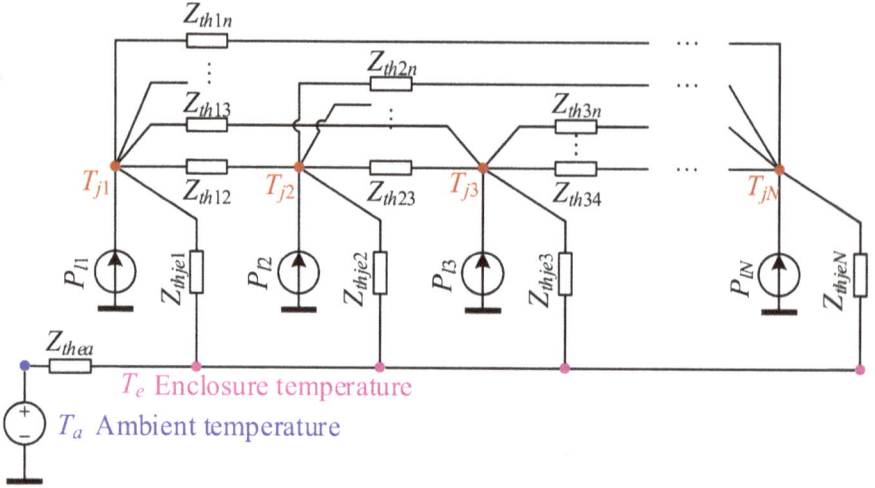

Figure 2.5   *Thermal impedance network of an enclosed converter system, including the self and mutual junction-enclosure thermal impedances*

the PV microinverter prototype has been built in ANSYS/Icepak, as shown in Figure 2.6.

Enclosure-ambient thermal impedance: The heat is transferred from the enclosure to the ambient by three methods, i.e., radiation, conduction, and convection. Meanwhile, the enclosure is a custom irregular cylinder, and thus, it is difficult to analytically quantify the enclosure-to-ambient thermal impedance Z_{thea}. Therefore, systematic FEM simulations have been conducted, and it is found that the aluminum enclosure is almost isothermal due to the filled high-thermal-conductivity compound. Based on the FEM simulations, the enclosure-ambient thermal impedance is modeled by Foster model:

$$Z_{thea} = R_{thea} \left(1 - e^{-t/(R_{thea}C_{thea})} \right) \tag{2.1}$$

where C_{thea} is a constant as $2\,673$ J/°C, but R_{thea} is determined by the total power loss P_l, i.e., $R_{thea} = 3.5P_l^{-0.216}$.

Junction-enclosure thermal impedance: Conduction is the primary heat-propagation approach inside the enclosed PV microinverter prototype, which implies that the system consisting of the enclosure, main components, and compound is linear and time-invariant. Thus, it is possible to apply the superposition principle. The junction/hotspot temperature of a device is thus derived as

$$
\begin{bmatrix}
T_{j1}(t) \\
T_{j2}(t) \\
\vdots \\
T_{jN}(t)
\end{bmatrix}
= \frac{d}{dt}
\begin{bmatrix}
Z_{je11}(t) & Z_{je12}(t) & \cdots & Z_{je1N}(t) \\
Z_{je21}(t) & Z_{je22}(t) & \cdots & Z_{je2N}(t) \\
\vdots & \vdots & \ddots & \vdots \\
Z_{jeN1}(t) & Z_{jeN2}(t) & \cdots & Z_{jeNN}(t)
\end{bmatrix}
*
\begin{bmatrix}
P_{l1}(t) \\
P_{l2}(t) \\
\vdots \\
P_{lN}(t)
\end{bmatrix}
+ T_e
\tag{2.2}
$$

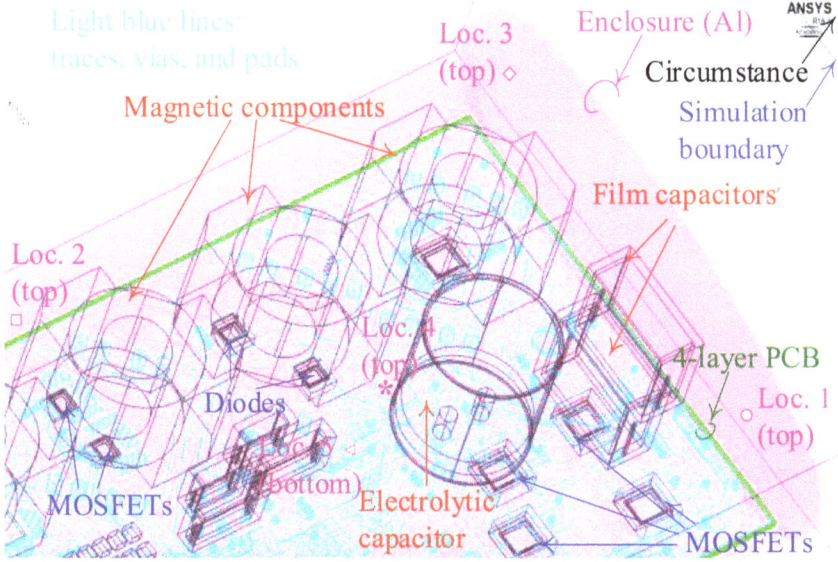

Figure 2.6   *Structure models of the main components, enclosure, and PCB (including traces and vias) built in ANSYS/Icepak for FEM simulations. The PCB and the enclosure are placed horizontally. The enclosure is naturally cooled, i.e., all faces are exposed to the open air.*

where T_{ji} is the junction/hotspot temperature of component i, Z_{jenn} represents the self junction/hotspot-to-enclosure thermal impedance, Z_{jemn} denotes the mutual junction/hotspot-to-enclosure thermal impedance between components m and n, T_e is the enclosure temperature, P_{ln} is the power loss of the nth component, and * denotes convolution.

Multiple FEM simulations are performed to extract the elements of the thermal impedance matrix in (2.2). Figure 2.7 shows the FEM simulation results for the self and mutual thermal impedances of S_1 and the self thermal impedances of other components. The FEM simulations for the junction-enclosure thermal impedances are then fitted as multiorder Foster thermal models, i.e.,

$$Z_{jemn} = \sum_{k=1}^{K} R_{jemn,k} \left(1 - e^{-t/\tau_{jemn,k}} \right) \qquad (2.3)$$

where Z_{jemn}, R_{jemn}, and τ_{jemn} are junction-enclosure thermal impedance, resistance, and time constant between components m and n, respectively.

The junction-enclosure temperature difference, enclosure-ambient temperature difference, and the junction temperature of component m can be calculated by the discrete equations

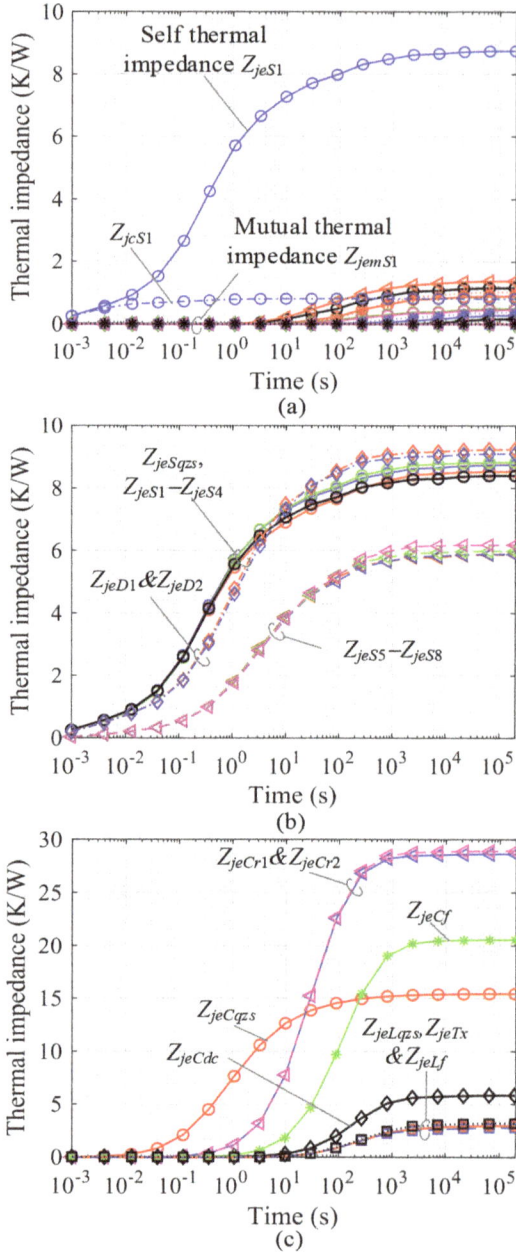

Figure 2.7   *FEM simulation results for thermal impedances. (a) Junction-case and junction enclosure thermal impedances of S_1; mutual junction-enclosure thermal impedances between S_1 and other components. Self junction-enclosure thermal impedances of (b) semiconductor devices and (c) passive components.*

$$\begin{cases} \Delta T_{jem} (x + 1) = \displaystyle\sum_{n=1}^{N} \sum_{k=1}^{K} \left[\begin{array}{l} \Delta T_{jemn,k} (x)\, e^{-t/\tau_{mn,k}} \\[4pt] + P_{l,n} (x)\, R_{mn,k} \left(1 - e^{-t/\tau_{mn,k}} \right) \end{array} \right] \\[12pt] \Delta T_{ea} (x + 1) = \Delta T_{ea} (x)\, e^{-t/\tau_{ea}} + P_{l,tot} (x)\, R_{ea} \left(1 - e^{-t/\tau_{ea}} \right) \\[8pt] T_{jm} (x + 1) = \Delta T_{jem} (x + 1) + \Delta T_{ea} (x + 1) + T_a \end{cases} \qquad (2.4)$$

where x denotes the time step, ΔT_{jem} is the junction-enclosure temperature difference, ΔT_{ea} represents the enclosure-ambient temperature difference, and T_{jm} is the junction temperature of component m.

2.2.3   Lifetime modeling

According to the failure mode and effect analysis results in [3, 4], the progressive increase of the on-state resistance (wear-out) of MOSFETs is mainly caused by the growth of fatigue cracks and voids into the source metal layer. A 20 percent rise of the on-state resistance is chosen as the criteria of wear-out failure, and a Coffin-Manson law-based reliability model is built in [3].

$$N_f = \alpha \cdot \left(\Delta T_j \right)^{-m} \qquad (2.5)$$

where N_f is the number of cycles of failure, ΔT_j is the junction temperature swing, and α and m are fitting parameters. A widely used capacitor lifetime model is employed for the lifetime projection of capacitors [5]:

$$L_{cn} = L_{c0} \times 2^{\frac{T_0 - T_h}{n_1}} \left(V/V_0 \right)^{-n_2} \qquad (2.6)$$

in which L_{cn} is the lifetime under the thermal and electrical stress T_h and V, L_{c0} is the lifetime under the reference temperature T_0 and the nominal voltage V_0. The coefficient n_1 is a temperature-dependent constant and n_2 is the voltage stress exponent. The values of n_1 and n_2 vary for different types of capacitors, and the relevant information is provided in [5].

A commonly used damage accumulation model—Miner's rule [6] is employed to calculate the accumulative damage of a component. A device fails when the accumulative damage, D_{mg}, reaches 1.

2.3   Wear-out failure analysis of the PV microinverter

The mission profiles from two locations, Arizona, USA, and Aalborg, Denmark, are applied to the PV microinverter models, yielding the annual temperature profiles of the critical components and enclosure, as shown in Figure 2.8. It can be seen that both the mission profile and the TCC effect have a significant impact on the junction/hotspot temperatures of components. When operating at Arizona and Aalborg (see Figure 2.8a and b), the highest component junction temperatures of the microinverter are 89 °C and 75 °C, respectively. If the TCC effect is neglected, a significant underestimation will be caused for the junction/hotspot temperatures, as shown in Figure 2.8c and d.

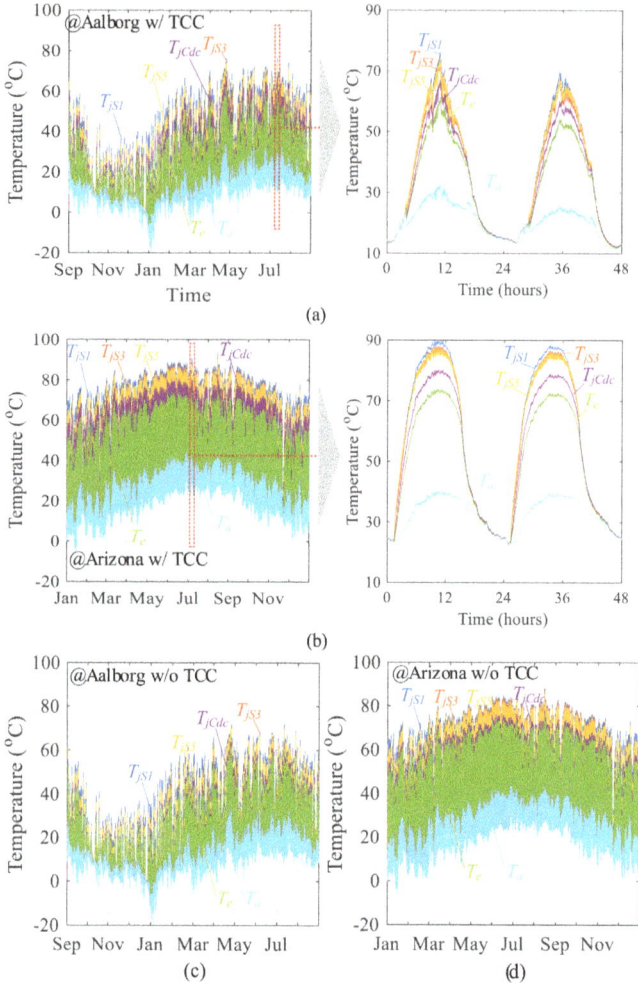

Figure 2.8   Temperature profiles of critical components (S$_1$, S$_3$, S$_5$, C$_{dc}$) and the enclosure. (a) Aalborg, Denmark, considering the TCC effect; (b) Arizona, USA, considering TCC; (c) Aalborg, Denmark, not considering TCC; (d) Arizona, USA, not considering TCC

2.3.1   Static annual damage of components

With and without considering the TCC effect, the annual damages of each critical component at the two locations are shown in Figure 2.9. It can be observed that the DC-link capacitor C_{dc} has the highest annual damage at both locations, i.e., 0.01 and 0.057 for Aalborg and Arizona, respectively. Assuming there are no other kinds of failures, the corresponding wear-out lifetimes of the DC-link capacitor are 100 years and 17.54 years for the two operating locations. However,

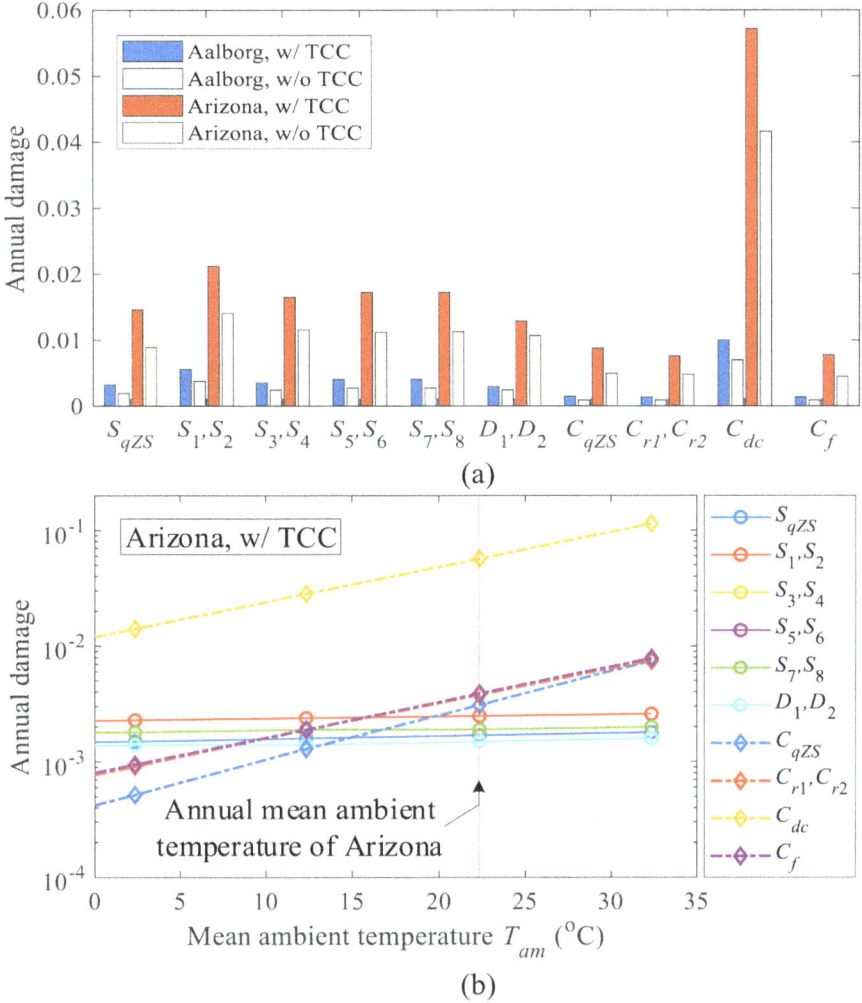

Figure 2.9   Annual damage to each critical component in the topology of
              Figure 2.1 when the microinverter operates at different locations with
              and without considering the TCC effect

if the TCC effect is not considered, then the annual damages of C_{dc} at the two locations are 0.007 and 0.031, which results in an underestimation rate of about 30 percent.

2.3.2   Monte Carlo simulation

In the real-world operation, many uncertainties, e.g., parameter boundaries of the employed device lifetime models, parameter variations of devices, and mission profile variations, will influence the lifespan projection of the PV microinverter.

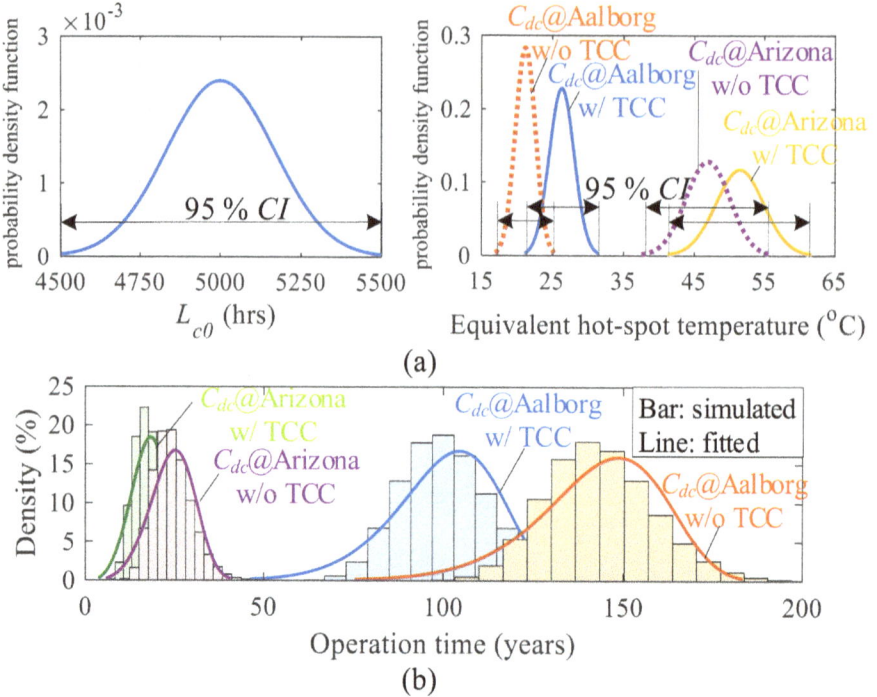

Figure 2.10   *Histograms of the years to the wear-out failure of (a) C_{dc} and (b) S_1 for a population of 1×10^5 samples operating at the two locations, with and without considering the TCC effect*

Considering all the variations that presumably obey the normal distribution, a sensitive analysis—Monte Carlo simulation is conducted. The histograms of years to wear-out failure for the selected components C_{dc} and S_s are shown in Figure 2.10a and b, respectively.

2.3.3   System failure probability due to wear-out

The histograms in Figure 2.10a and b are fitted as the Weibull distribution [7]. Assume all the considered devices are connected in series in the reliability model, i.e., any component failure will lead to system failure. Then the system wear-out failure $F_{sys}(t)$ is

$$F_{sys}(t) = 1 - \prod (1 - F_i(t)) \tag{2.7}$$

where $F_i(t)$ represents the cumulative distribution function of wear-out failure of the ith component.

Figure 2.11 shows the probability curves of wear-out failures for components and the system when operating at Aalborg, Denmark, and Arizona, USA, with and without considering the TCC effect. First, it can be seen that the mission

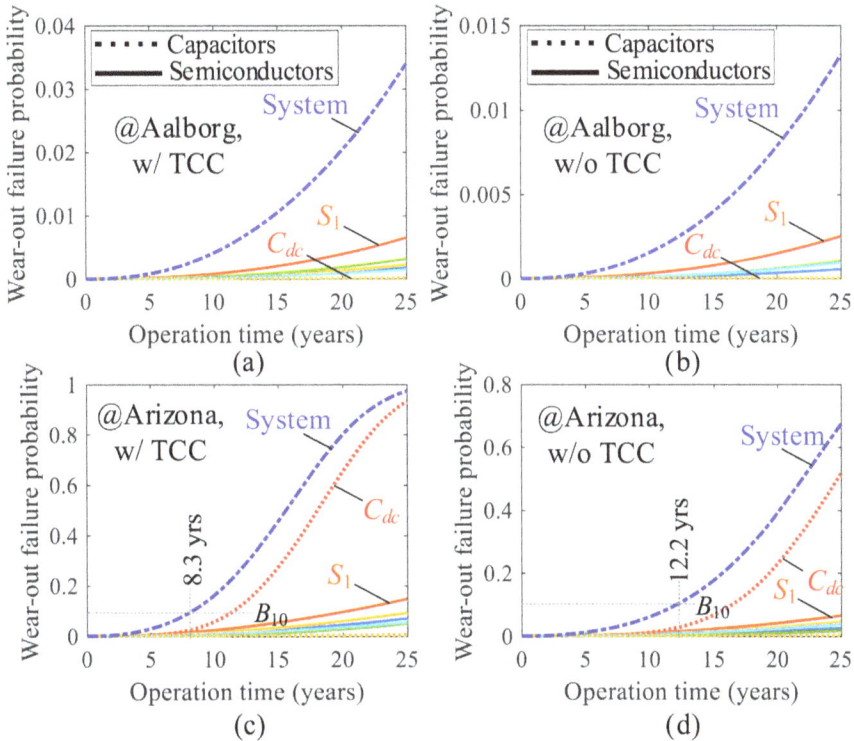

Figure 2.11   *Probability curves of wear-out failure for each component and the system when operating at (a) Aalborg, Denmark, with considering the TCC effect; (b) Aalborg, Denmark, without considering the TCC effect o; (c) Arizona, USA, with considering the TCC effect; (d) Arizona, USA, without considering the TCC effect*

profile has a strong impact on the wear-out failure: when operating in a harsher environment, i.e., Arizona, the wear-out failure probabilities before 25-year operation are significantly higher. Second, neglecting the TCC effect will lead to an obvious underestimation of the wear-out failure probability; when operating at Aalborg, the predicted system wear-out failure probability before 25 years is 3.3 percent (see Figure 2.11a), whereas the corresponding value is only 1.3 percent (see Figure 2.11b) if neglecting the TCC effect. When operating at Arizona, the B_{10} lifetimes with and without considering the TCC effect are 8.3 years and 12.2 years (see Figure 2.11c and d), respectively, which implies that about 45 percent lifetime overestimation can be made if the TCC effect is neglected. Also, it can be seen that the DC-link electrolytic capacitor C_{dc} has the highest wear-out failure probability when the operating environment is harsh and thus dominates the system wear-out failure. Hence, it can be concluded that the DC-link electrolytic

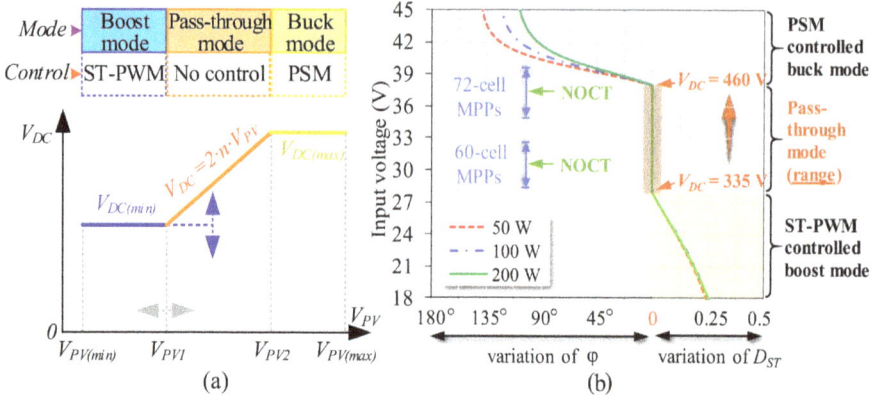

Figure 2.12   Advanced multimode control of the qZSSRC with a variable DC-link voltage: (a) sketch of DC-link voltage variations; (b) regulation characteristics. The abbreviations "PSM", "ST", "PWM", "MPP" and "NOCT" represent "phase-shift modulation", "shoot-through", "pulse-width modulation", "maximum power point" and "nominal operating cell temperature", respectively.

capacitor C_{dc} is the bottleneck of 25-year reliable operation for the studied PV microinverter.

2.4   Reliability improvement of the PV microinverter

2.4.1   Advanced multimode control of the qZSSRC

With the conventional multimode control scheme proposed in [1], the DC–DC converter operates in the PTM only at a particular input voltage, in which the converter has the highest efficiency, as shown in Figure 2.3e. However, it is not necessary for a grid-connected microinverter to have a stable DC-link voltage; hence, an advanced multimode (variable DC-link voltage) control scheme (see Figure 2.12a) can be applied to the DC–DC converter, as illustrated in Figure 2.12b. Specifically, the operation range of the PTM is $V_{in} \in$ [28, 38] V, and in this mode, the DC-link voltage varies with respect to V_{in} from 335 V to 460 V. When V_{in} is beyond the PTM range, the microinverter will operate in the buck or boost mode, and the DC-link voltage will be fixed at 460 V and 335 V, respectively. With the advanced control scheme, the efficiency performance of the microinverter can be significantly improved, as shown in Figure 2.13. Thus, the power losses and junction/hotspot temperatures of components can be remarkably decreased.

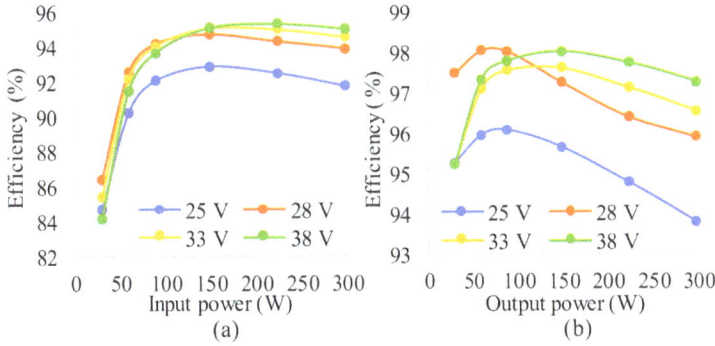

Figure 2.13 *Measured efficiency with the new control strategy: (a) the whole PV microinverter including the auxiliary power, (b) the DC–DC power stage*

2.4.2 New DC-link electrolytic capacitor with longer nominal lifetime

It has been observed from Figure 2.11c and d that the used DC-link electrolytic capacitor C_{dc} is the bottleneck for the long-term reliable operation of the commercial PV microinverter product. In the baseline design, the DC-link employs a cost-optimized 150-μF electrolytic capacitor, whose nominal lifetime is 5 000 hours at 85 °C. To improve system reliability, this chapter proposes to replace

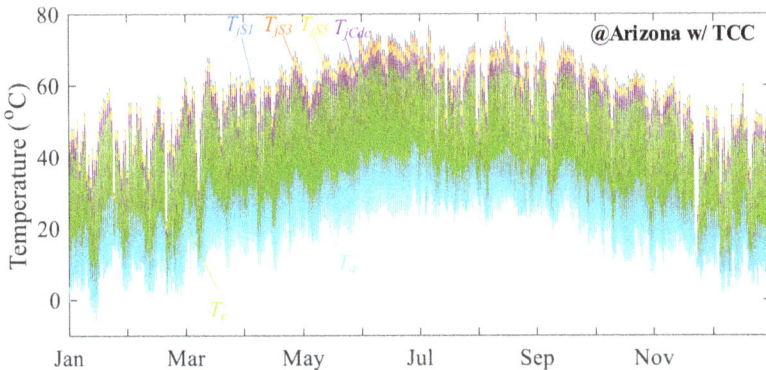

Figure 2.14 *Calculated temperature profiles of critical components (S_1, S_3, S_5, C_{dc}) and the enclosure of the PV microinverter with the new DC-link capacitor and new control scheme. The mission profile of Arizona is used and the TCC effect is taken into account in the temperature calculations.*

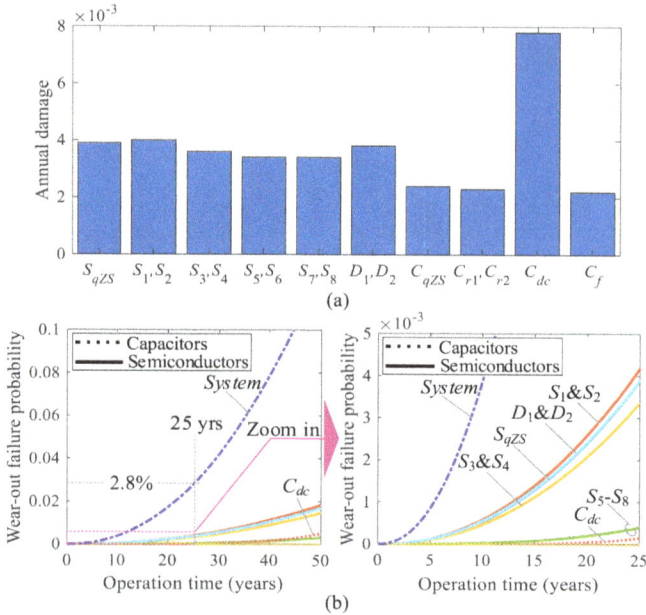

Figure 2.15   Reliability evaluation results of the PV microinverter with the variable DC-link voltage control and the new electrolytic capacitor; the mission profile of Arizona is applied: (a) annual damage and (b) wear-out failure probabilities of each component and the system

the previous electrolytic capacitor with a new one with a longer nominal lifetime (5 000 hours @105 °C) along with the advanced multimode control scheme.

2.4.3   Wear-out failure probability

With the proposed advanced multimode control and the better electrolytic capacitor, the temperature profiles of S_1, S_3, S_5, C_{dc}, and the enclosure are derived, as shown in Figure 2.14. Compared with the baseline solution, the new design enables the microinverter to operate at lower temperatures (see Figures 2.8 and 2.14). Figure 2.15 shows the obtained annual damage and wear-out failure probability with the new design.

Because of the lower junction/hotspot temperature, the annual damages of components are reduced; notably, the annual damage of C_{dc} is remarkably decreased from 0.057 to 0.0078. The wear-out failure probabilities of the components and the system are shown in Figure 2.15b. With the new design, all the failure probabilities before 25 years are maintained low. For the system, its wear-out failure probability over 25-year operation is about 2.8 percent, which is significantly enhanced compared with the baseline solution (see Figure 2.9c).

2.5 Summary

This chapter presents a wear-out failure analysis and reliability improvement of a 300-W impedance-source PV microinverter product. A description of the product configuration and reliability evaluation process is first presented. With experimental measurements and FEM simulations, a system-level electrothermal model and empirical lifetime models are then built for the components and enclosure of the PV microinverter. After that, the wear-out performance of the microinverter is evaluated before measures are taken to improve its reliability. It is concluded that (i) both the mission profile and TCC effect have a remarkable impact on the lifetime projection of the PV microinverter; (ii) the DC-link capacitor is the weakest link in the product in terms of the wear-out performance; and (iii) the proposed variable DC-link voltage control and long-lifetime aluminum electrolytic capacitor can significantly reduce the system wear-out failure probability.

References

[1] Vinnikov D., Chub A., Liivik E., Roasto I. 'High-performance quasi-Z-source series resonant DC–DC converter for photovoltaic module-level power electronics applications'. *IEEE Transactions on Power Electronics*. 2017;**32**(5):3634–50.

[2] TamizhMani G. 'Standardization and reliability testing of module-level power electronics MLPE'. *NREL Reliability Workshop*. 2015:1–18.

[3] Testa A., De Caro S., Russo S. 'A reliability model for power MOSFETs working in avalanche mode based on an experimental temperature distribution analysis'. *IEEE Transactions on Power Electronics*. 2012;**27**(6):3093–100.

[4] Russo S., Testa A., De Caro S., *et al.* 'Reliability assessment of power MOSFETs working in avalanche mode based on a thermal strain direct measurement approach'. *IEEE Transactions on Industry Applications*. 2016;**52**(2):1688–97.

[5] Shen Y., Chub A., Wang H., Vinnikov D., Liivik E., Blaabjerg F. 'Wear-out failure analysis of an impedance-Source PV microinverter based on system-level electrothermal modeling'. *IEEE Transactions on Industrial Electronics*. 2019;**66**(5):3914–27.

[6] Miner M.A. 'Cumulative damage in fatigue'. *Journal of Applied Mechanics*. 1945;**12**(3):A159–64.

[7] McPherson J.W. *Reliability Physics and Engineering*. 2nd edn. Berlin/Heidelberg: Springer; 2013.

Chapter 3

Reliability analysis methods and tools

Ionut Vernica[1] and Frede Blaabjerg[1]

3.1 Background and motivation

Nowadays, due to the growing concerns regarding the impact of greenhouse gas emissions on the environment, renewable energies are seen as part of the solution to this problem and are gaining much of the attention of both public and private stakeholders. Among renewable energy solutions, photovoltaic (PV) technology can provide a large share of the future clean energy demands. Enabled by the advancements in semiconductor technologies, the cost of PV systems has seen a significant drop in recent years [1] and, thus, facilitating a rapid growth in the PV installation capacity over the past decade [2].

With increased penetration level of PV systems, the safety and reliability requirements of such systems have also become more stringent over time. For example, currently, a typical PV system is expected to have a target lifetime between 5 and 30 years, considering 12 hours per day operation [3]. These reliability requirements are likely to tighten in the near future, and PV systems will be expected to withstand more harsh environments and longer operating hours, with failure return rates in the parts per million (ppm) range [4].

Within the PV system, the inverter is an essential subsystem that can have a significant effect on the overall volume, efficiency, availability, and reliability of the entire system. However, by looking at the quantified reliability metrics (e.g., operation and maintenance events) reported by different PV plant operators [5–7], it can be noticed that the inverter represents one of the most prone-to-failure subassemblies of the PV system. Despite its crucial role in the correct operation of the overall system, as shown in Figure 3.1, the failure rate of the PV inverter can vary between 43 and 70 percent in commercial, residential, and utility PV installations. Thus, the inverter is often seen as the "bottleneck" of the PV system, with respect to reliability. Furthermore, as shown in Figure 3.2, from the cost breakdown of unscheduled maintenance events that occurred during the five-year operation of a utility-scale PV generation plant [8], it can be concluded that the PV inverter is the highest source of repair and maintenance cost in the PV system. Considering the additional costs associated with energy losses due to

[1]Department of Energy Technology, Aalborg University, Aalborg, Denmark

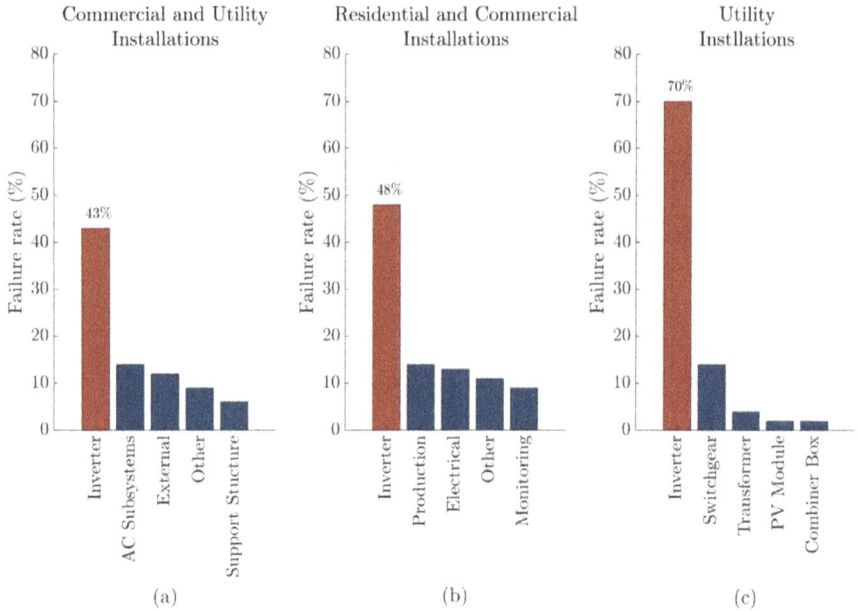

Figure 3.1   Failure rate distribution for (a) commercial and utility PV installations [4], (b) residential and commercial PV installations [5], and (c) utility PV installations [6]

downtime, the reliability of the PV inverter can have a significant impact of the levelized cost of energy (LCOE).

Consequently, in an attempt to improve the reliability performance of the PV inverter and inherently reduce the number of unexpected maintenance events of the PV system, various reliability assessment methods and tools are employed. Ideally, the reliability analysis and allocation should be performed during the design and development stages of the product, as it is much more cost-effective to identify major flaws and design weaknesses in the PV inverter early on in the product life cycle. Additionally, considering that the design and manufacturing processes have been found responsible for more than half of the inverter failures [9], the reliability growth strategy should be built around employing reliability evaluation tools during these two crucial stages of the product life cycle.

As shown in Figure 3.3, several methods and tools can be used to assess and improve the reliability of PV inverters. However, the applicability and cost-effectiveness of each tool are strongly dependent on the product life cycle stage. A general overview of the most common reliability tools used for PV inverters is given in this chapter. The presented tools can be employed during both the early stages of the product development process (e.g., failure mode and effect analysis (FMEA), Design for Reliability (DfR), and custom or commercial software tools) and the more mature stages of the product life cycle (e.g., reliability testing and robustness validation). To provide the reader with a better understanding, practical examples on an Insulated Gate Bipolar Transistor (IGBT) inverter power module are given where possible.

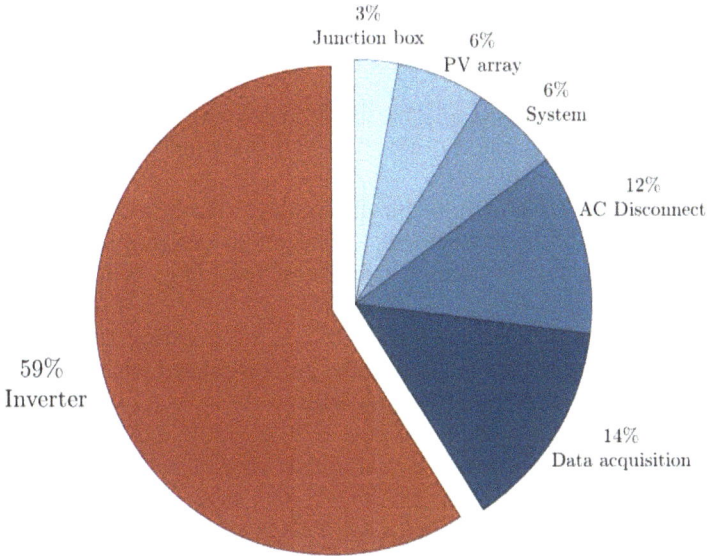

Figure 3.2   Unscheduled maintenance cost breakdown of a utility-scale PV installation [8]

3.2   Failure mode and effect analysis

FMEA is a systematic methodology used to identify and understand the possible failure modes (FMs) of a system or process and assess their associated causes and effects to justly determine (and prioritize) corrective or preventive measures. Due

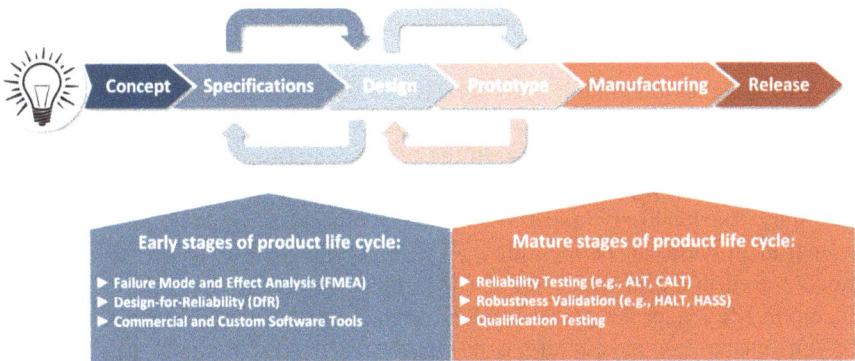

Figure 3.3   Different reliability analysis methods and tools, and their applicability throughout the various stages of the product life cycle

to its versatility, FMEA can be applied throughout different stages of the product development process, and, thus, it can be classified into three major types:

- **System FMEA (SFMEA)**—Typically employed during the early concept stages of the product, when a system-level analysis is possible and required (e.g., PV inverter system).
- **Design FMEA (DFMEA)**—Typically employed during the design (and development) stages of the product with a focus on the design-related issues, which may appear at the subsystem or component level (e.g., IGBT power module used in the PV inverter).
- **Process FMEA (PFMEA)**—Typically employed during the manufacturing stage of the product and focuses on the process-related issues associated with the manufacturing and assembly processes.

In general, FMEA should be performed when a new system or process is being developed, or when an existing system or process suffers significant modifications with high associated risks. FMEA relies strongly on knowledge regarding the design of the system or process, its mission profiles, and possible historical data from similar systems or processes (e.g., field failure statistics). Thus, to effectively use this tool and link the possible FMs to the potential causes and effects of its critical subsystems or components, prior information concerning the application use-cases is needed. Otherwise, in cases where there is a lack of available information or historical data (e.g., as is the case for new technologies/processes), the effectiveness of the FMEA depends solely on the knowledge and experience of the team carrying out the analysis. A general FMEA methodology is proposed in the IEC 60812 standard [10], whereas broader FMEA implementation guidelines can be found in [11].

In this chapter, the applicability of the DFMEA procedure is exemplified on the IGBT power module component of the PV inverter. The analysis helps to identify the possible FMs of the power module and to understand its causes and effects on the overall PV system (e.g., cease or significantly degrade the operation, affect the safety of the user). To perform DFMEA in a systematic manner, several graphical tools can be employed.

The boundary diagram is used to represent the physical relationships between the components of the system and, as the name suggests, the boundary of the analysis. The interdependencies between the target component or subsystem under a specific life cycle stage and its interfacing elements can be graphically illustrated, and, thus, the functions (and possible interferences) of the component or subsystem of interest can be identified. A sample "boundary diagram" for a power module used in a grid-connected PV inverter is shown in Figure 3.4. During its operation life cycle, the power module interacts with various components (e.g., DC-link busbar, printed circuit board assembly (PCBA) solder joint, cables) of the PV system. As shown in Figure 3.4, the interactions between the power module and each of the interface elements can be visually defined. For example, the power module interacts with the DC-link busbar through the current and harmonics it

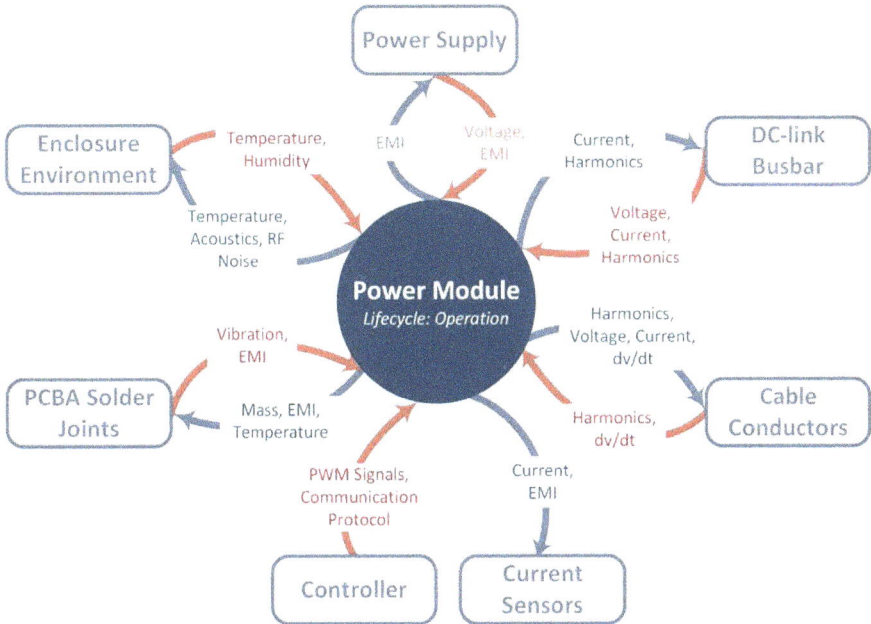

Figure 3.4   *"Boundary diagram" example of a power module used in a typical grid-connected PV inverter during the field operation life cycle. (PCBA = printed circuit board assembly, dv/dt = rate of change of voltage, EMI = electromagnetic interference, PWM = pulse width modulation, RF = radio frequency).*

generates, and through the voltage, current, and harmonics it receives from the DC-link busbar.

Another useful graphical tool that can facilitate the DFMEA process is the "Parameter Diagram" (P-Diagram). This tool illustrates the ideal correlation between the input of the component or subsystem and its desired output response. Additionally, the sensitivity of the specified component or subsystem function to external noise and control factors is considered, and potential error states (or unintended outputs) can be defined. Control factors are typically the adjustable design parameters of the component or subsystem, while noise factors represent the factors that can indirectly influence the design and behavior of the component or subsystem (e.g., piece-to-piece variations, degradation, customer usage profile, environmental or operating conditions, and system interactions). Finally, the error states can be defined as any kind of loss in functionality or any other unintended outputs. An example of a "P-Diagram" for the power module of a grid-connected PV inverter is shown in Figure 3.5. The inputs required by the power module to correctly perform its designated function and its ideal outputs are highlighted. To convert the DC electrical energy to controlled AC electrical energy and inherently generate controlled AC voltage and current signals, the power module requires a

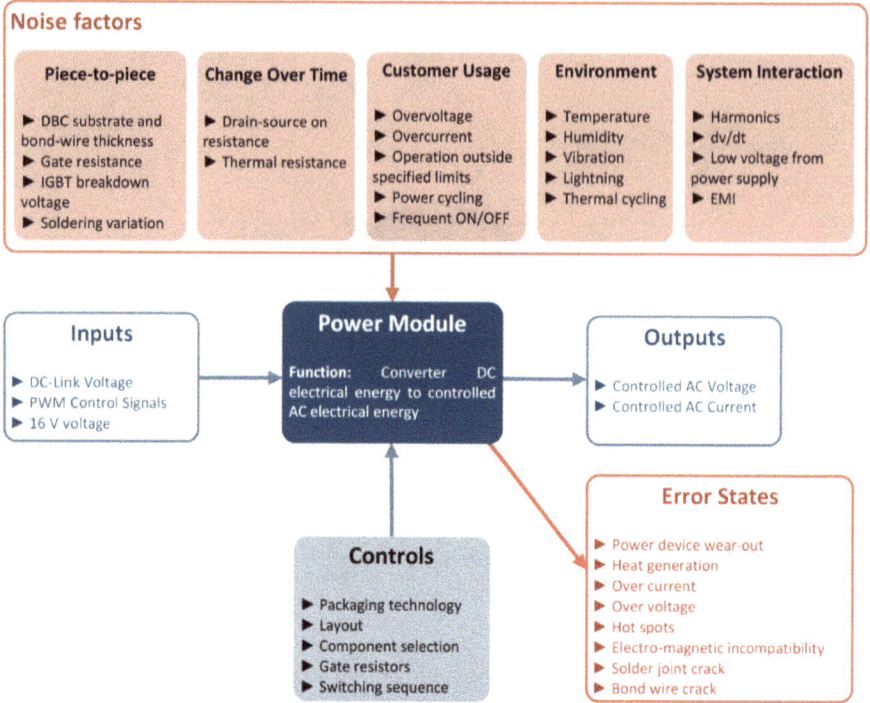

Figure 3.5   Sample "P-DIagram" Representation of a Power Module Used Within a Typical Grid-Connected PV Inverter. (PWM = Pulse Width Modulation, DC = Direct Current, AC = Alternative Current, DBC = Direct Bonded Copper, IGBT = Insulated-Gate Bipolar Transistor, DV/DT = Rate of Change of Voltage, EMI = Electromagnetic Interference)

DC voltage, a sequence of PWM control signals, and a 16-V power supply voltage as inputs. The functionality of the power module can be influenced by several control factors, such as packaging technology, layout configuration, which will, in turn, result in variations in the AC current and voltage outputs. Moreover, external noise factors can lead to a suboptimal operation of the power module and several error states, such as wear-out, excessive heat generation, and bond-wire crack.

Finally, the "structure tree" tool can provide a graphical representation of the system structure and identify the interactions between the several subsystems or components of the PV systems. The top-down approach is typically the most effective, giving an overview from product level down to subcomponent level. An example structure tree of a typical PV system is given in Figure 3.6. As shown in Figure 3.6, the PV system can be divided into three major independent systems (e.g., PV module, PV inverter, and Balance of System (BOS)), which can then be further classified into different subsystems. The IGBT power module lies at the component level, under the PCBA subsystem, and with several subcomponents at its bottom

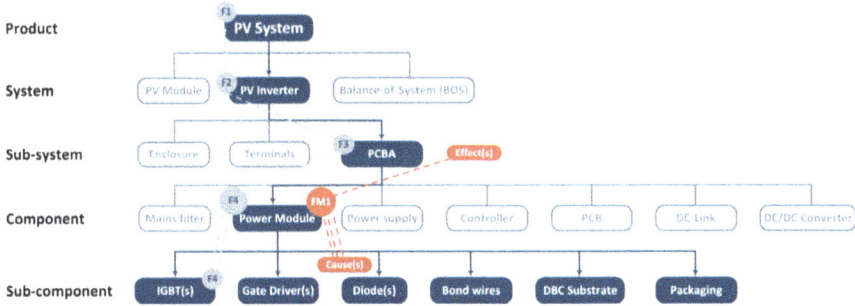

Figure 3.6   *Example "structure tree" of a power module used in a typical PV inverter of a grid-connected PV system. (PCBA = printed circuit board assembly, PCB = printed circuit board, IGBT = Insulated-Gate Bipolar Transistor, DBC = direct bonded copper, FMx = failure mode, Fx = function).*

level. Additionally, the "structure tree" can provide the designer with an optimal tool on which the FM and functional decomposition can be performed. Sample functions (F) and their corresponding FMs are given below for the power module component of the PV inverter.

F4.1: Convert DC electrical energy to controlled AC electrical energy
FM-01: IGBTs do not conduct when turned ON;
FM-02: IGBTs do not block when turned OFF;
…
FM-*N*: No commutation signal from gate drivers;
F4.2: Connect to DC busbar and cable conductors
FM-01: Weak connection;
FM-02: Crack in solder joints;
…
FM-*N*: Connection in wrong position;
F4.3: Withstand environmental conditions
FM-01: Subcomponent failure due to excessive humidity;
FM-02: Subcomponent failure due to excessive ambient temperature;
…
FM-*N*: Subcomponent failure due to excessive air pollution;
…
F4.N: Operate below electrical loss target
FM-01: Too high switching losses;
FM-02: Low-quality electrical input (e.g., harmonics)
…
FM-*N*: Degradation of internal parameters (e.g., thermal resistance)

Each component or subsystem of the PV system can perform multiple functions, and each function can have several FMs that can interrupt or degrade its functionality. Furthermore, each FM has a "cause," which represents the specific reason for the failure, and an "effect," which is the consequence of the failure. Typically, the causes can be linked to the below-layer of the "structure tree," and can be tracked down until the root cause is identified. The effect of a specific FM is typically visible in the above-layer of the "structure tree."

To determine the impact of different FMs on the system and to correctly define and assign preventive or corrective measures, each FM can be rated according to its severity of effect (S), likelihood of occurance (O), and likelihood of detection (D). Thus, based on the above-mentioned factors, the Risk Priority Number (RPN) can be computed and used as a numerical ranking system of each potential FM. The RPN can be calculated as shown below:

$$RPN = S \cdot O \cdot D \tag{3.1}$$

Thus, based on the RPN ranking, educated decisions or actions can be taken to reduce the likelihood of occurrence of a certain FM (or improve its detectability), and, hence, the impact of a given FM can be minimized (or even mitigated) and the reliability of the PV system can be improved. Generic scales for the severity, occurrence, and detectability ratings can be found in [12]. However, typically, these factors need to be adapted in accordance with the requirements of PV plant operator or user.

In conclusion, FMEA is an effective and flexible tool, which can help the designer to identify potential FMs early on in the development process of PV inverters and can help determine proper design methods for a sustainable reliability growth strategy. However, for complex systems with many subsystem or components, the analysis can become very time-consuming and resource-demanding. Additionally, the lack of available product use-cases and historical data can represent a significant drawback as the analysis is based solely on the experience and knowledge of the design team (e.g., a multidisciplinary team is needed).

3.3   Design for reliability

Another powerful reliability assessment and improvement tool, which can be used during the early life cycle stages of PV inverters, is the DfR methodology. The concept of DfR is based on the processes performed during the early design and development stages of a product, which guarantee that certain reliability or quality design targets are met [3]. An effective implementation of the DfR method can help the designer to quickly benchmark different technologies, accelerate the reliability processing time, and, inherently, reduce the time-to-market of new products and their overall life cycle cost. This has led to the DfR concept to be widely adopted since the early 1950s. However, for PV inverters, the methodology has been gaining significant attention in recent years due to the progress in software simulation tools and increased testing capabilities of power electronic components.

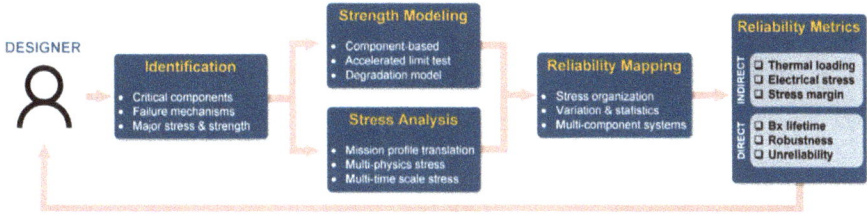

Figure 3.7   Generic Flow Diagram of the DfR Methodology Used for Power-Electronic-Based Systems

A generic flow diagram of the DfR procedure is given in Figure 3.7. The analysis begins with the *Identification* phase, during which the designer classifies the critical components of the system and their corresponding FMs and failure mechanisms. This information can be obtained through other reliability analysis tools and concepts, such as FMEA and Physics-of-Failure (PoF). Moreover, during the first phase of the DfR approach, the stress and strength profile of the critical components can be determined with the help of robustness validation testing (e.g., highly accelerated life testing (HALT)). Within the second stage of the process, the critical component information can be used by the designer for *Strength Modeling* and *Stress Analysis*. Typically, the strength modeling is performed at component-level and reliability testing methods, such as Accelerated Life Testing (ALT), are used to determine an analytical lifetime model equation, which can describe a particular wear-out failure mechanism. On the other hand, within the *Stress Analysis* part, the environmental and operating mission profiles of the system are identified and used to analytically calculate the corresponding stressors of the components of interest, through specialized simulation software tools. Following, based on the developed strength and stress models, the *Reliability Mapping* phase of the DfR approach can commence. During this phase, the stress data needs to be organized and represented in such a way that it can be correctly applied to the lifetime model and determine the initial lifetime estimation of the components. For a more realistic approach, possible variations are included and different statistical tools (e.g., Weibull, Monte Carlo simulation) are used to determine the lifetime distribution of the components of interest. System-level reliability tools (e.g., Reliability Block Diagram (RBD)) can then be used to calculate the reliability performance of multicomponent systems. Finally, throughout the DfR procedure, the designer can obtain several critical reliability metrics, either as indirect reliability indicators (e.g., expected electrical and thermal loading or behavior) or as direct reliability indicators (e.g., estimated lifetime distribution, unreliability curve).

Thus, by applying the DfR concept, the designer can use the obtained reliability indicators to identify the major design weaknesses and flaws, and take effective decisions or actions regarding the reliability-oriented design of the product. Consequently, motivated by the potential cost-saving and time-to-market reduction benefits of the DfR methodology, mission-profile-based reliability assessment procedures for power electronics have been proposed throughout the scientific literature

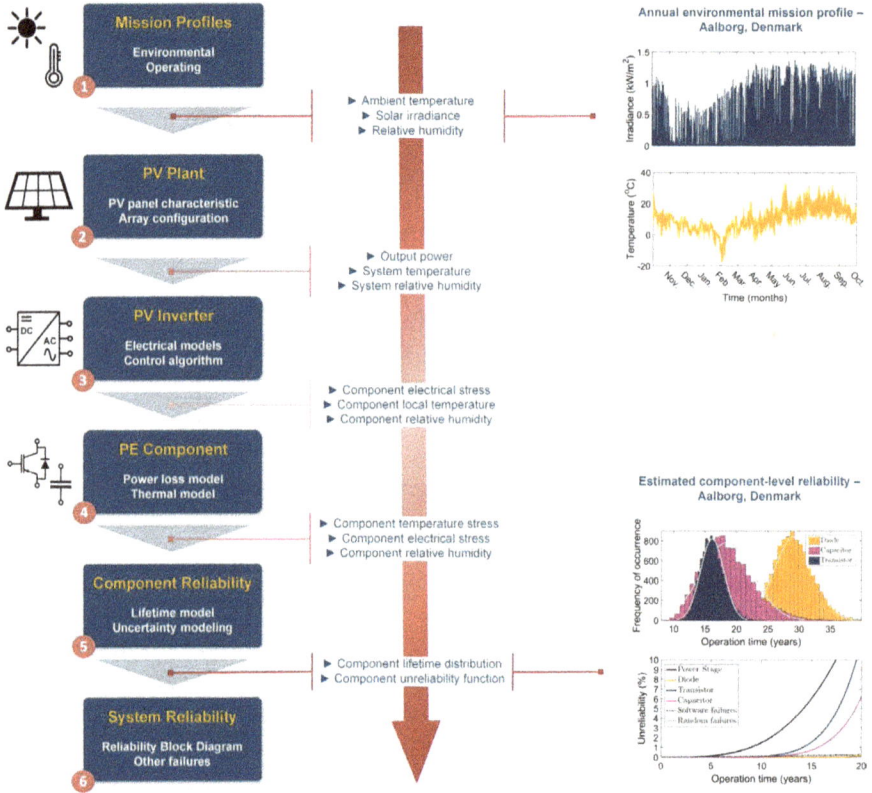

Figure 3.8   Typical Mission-Profile-Based Lifetime Estimation Methodology for Power Electronics Used in PV Systems. (PE = Power Electronic)

for a wide variety of applications (e.g., wind power converters [13], more electric aircrafts [14], traction inverters for electric vehicles [15]). Among these, mission-profile-based lifetime estimation methods of PV inverter systems have been the focus of many studies [16–18].

A general mission-profile-based reliability assessment procedure of power electronics used in PV systems is presented in Figure 3.8. The presented six-step model-based approach is used to translate the environmental mission profiles to the system-level, converter-level, and then to the component-level stress, and, finally, to estimate the component-level and the system-level reliability metrics. This methodology integrates the various physical layers (e.g., electrical, thermal, lifetime, statistical) of the reliability analysis process in a straightforward manner, thus, allowing for a quick adoption and implementation by either PV inverter manufacturers, users, or integrators. A brief description of each of the six steps is given below:

- **Step 1: Mission Profile**—Initially, the designer must determine what are the relevant operating and environmental conditions to which the PV systems are exposed to. The cumulation of these conditions represents the mission profile of the system. Typically, field-measured data of ambient temperature, solar irradiance, and relative humidity represent a generic mission profile for PV systems.

- *Step 2: PV Panel(s)*—During this step, the input mission profiles are converted to the system-level outputs (e.g., system-level temperature or relative humidity and output electrical power of the PV panel). This can be achieved with the help of electromechanical analytical models of the PV panel or through look-up table approximations.
- *Step 3: PV Inverter*—Based on the analytical electrical models of the power electronic system, the output electrical power of the PV panel is used to determine the electrical stresses (e.g., voltage or current loading) of each of the components of interest (e.g., IGBT, capacitor). The impact of the control algorithm and/or of the modulation technique can also be included within this step. Moreover, the system-level temperature or relative humidity can be translated to the local temperature or relative humidity of each component, either through rough "rule-of-thumb" approximations or through complex Finite Element Analysis (FEA) simulations.
- *Step 4: Power Electronic (PE) Component*—In the next step of the procedure, the electrical loadings of the components of interest are translated into their corresponding thermal dynamic behavior. This is achieved according to the analytical electrothermal models, which are used to calculate the generated power losses of the components and, inherently, their thermal loadings. Additional information regarding the external thermal network (e.g., cooling) of the components can be included within this step for a more accurate estimation of the thermal stress.
- *Step 5: Component Reliability*—Within this step, the previously determined stressors of the components of interest (e.g., temperature, electrical and/or relative humidity) are linked to their corresponding lifetime models and, thus, used to estimate the expected lifetime of the component. For a more realistic approach, the uncertainties that occur in the stressors or lifetime model coefficients are included in the statistical analysis, and different reliability metrics (e.g., lifetime distribution, unreliability curve) are identified.
- *Step 6: System Reliability*—Finally, the reliability information of each of the components of interest can be merged together through various methods (e.g., RBD, Markov chain), and the system-level reliability can be estimated. Additionally, at this point, other FMs, which are difficult to model or calculate (e.g., catastrophic failures, software failures), can be approximated (e.g., annual percentage) and included in the final reliability estimation.

The above-mentioned mission-profile-based reliability assessment procedure can aid the designer to quickly identify major design flaws and implement preventive or corrective measures (e.g., redundancy, overdesign of fragile components), which can help to improve the design of the PV inverter with respect to reliability. If accurate data or information about the system are available to the designer, then the procedure can be effectively used to determine the product warranty periods and service intervals or schedule predictive maintenance routines. However, for cases where there is a lack of accurate available data, the given procedure could still facilitate the DfR methodology by allowing for fast benchmarking of different solutions and technologies, and quantifying their impact on the relative lifetime prediction.

In conclusion, the DfR methodology is a powerful asset in the designer's toolbox, which can be used during the early design and development stages of the PV inverter life cycle. A proper implementation can result in significant advantages, such as reduced time-to-market, reduced overall life cycle cost, or as a means of rapid comparison of the impact that different technologies have on the component or system reliability. However, an effective implementation of the DfR methodology requires it to be used together with other reliability analysis methods and tools, such as FMEA, fault tree analysis, simulation software, and accelerated life testing.

3.4 Software tools in design for reliability

Due to the rapid growth of available computational power and capabilities of general-purpose computers, simulation tools have become much more efficient in recent years and, thus, have come to be one of the most wide-spread tools among engineers. Based on defined input conditions, the design engineer can simulate the dynamic behavior of the component or system and get an accurate overview of its performance. As a result, potential design flaws can be identified before a physical prototype is built and, thus, saving valuable cost and time resources. However, an "all-purpose" simulation tool does not exist, and typically, different tools have various focus areas and intended purposes. As an example, computer-aided design tools are used to model mechanical structures, layout configurations, etc., which can then be used by specialized FEA software for different types of simulations (e.g., mechanical, electromagnetic interference, thermal).

For PV inverters, by far the most adopted and used tools by engineers are FEA software and circuit simulators. FEA tools such as ANSYS® [19] or COMSOL® [20] provide advanced numerical methods for multiphysics simulations, which can help the designer to perform complex analyses regarding the electrical response of the circuit, electromagnetic compatibility/interference, dynamic thermal behavior, or even the component or system-level reliability performance at both microelectronics and power electronics level. On the other hand, circuit simulators such as MATLAB Simulink® [21], PLECS® [22], or SaberRD® [23] are used extensively to simulate the electrical and (in some cases) thermal behavior of the PV inverter power electronics and microelectronics circuits, together with their corresponding auxiliary interface elements (e.g., control, grid). At this point, it should be noted that for many simulation tools, microelectronics and power electronics are different focus areas, as they exhibit different standardization levels, characteristics, and/or dynamic behavior models. The presented tools can be used during the early stages of the PV inverter development process to determine its indirect reliability metrics according to some predefined input conditions and assumptions. The electrical behavior of the system, the generated power losses, or the expected heating temperature of the components are just some of the indicators that can be used to estimate the reliability of the PV inverter and, inherently, optimize its design in terms of specific electrical, thermal, or lifetime requirements.

However, when considering reliability-focused simulation tools for PV inverters, not many options are available in the market. Despite their strong capabilities of estimating both component and system-level reliability of microelectronic components, the reliability-oriented features of power electronics implemented in ANSYS are still quite limited. Similarly, MATLAB Simulink and PLECS do not include specific reliability-oriented features for PV inverter applications but allow for certain custom add-ons to be developed and used to extend their use-case to the reliability of power electronics. Consequently, due to the limited reliability functions of circuit simulators and FEA software, general-purpose reliability software is typically employed. One of the most widely used statistics-based reliability software is ReliaSoft® [24]. Its multitude of toolboxes allow the design engineer to perform a wide variety of analyses, such as reliability prediction and allocation, B_x lifetime characterization, system-level reliability modeling, or FMEA. All the previously mentioned features make ReliaSoft a powerful tool, which can be used to translate field-failure data (or test data) to relevant direct reliability metrics, such as unreliability curve or lifetime distribution.

In addition to the above-mentioned commercial solutions, custom software tool platforms developed by academic or research institutions or by PV component or system manufacturers can also be used for estimating the direct and/or indirect reliability metrics of the PV inverter. For example, PowerSynth is a multi-objective optimization software tool developed at the University of Arkansas, USA, which allows for rapid design and verification of IGBT power modules by means of reduced-order modeling [25]. The Simulation Assisted Reliability Assessment (SARA®) software tool has been developed at the Center for Advanced Life Cycle Engineering (CALCE), University of Maryland, USA, and uses PoF-based principles to assist its users in the lifetime assessment and reliability estimation of microelectronic components and subassemblies [26]. Another custom software tool platform that can be employed during the design and development stages of PV inverters is the Design for Reliability and Robustness (DfR²) tool [27, 28], developed at the Center of Reliable Power Electronics (CORPE), Aalborg University, Denmark. The DfR² tool allows for rapid and straightforward estimation of the power electronic component or system reliability, under realistic long-term environmental mission profiles and operating conditions. Finally, online simulation tools provided by PV component or system manufacturers are worth mentioning, as they provide an accessible way of estimating the indirect or direct reliability metrics of a specific component or system, based on its datasheet parameters and operating conditions. For example, tools such as SemiSel® [29] and IPOSIM® [30] are provided by IGBT power module manufacturers and can be used to calculate indirect reliability indicators such as generated power losses or dynamic thermal behavior. Thus, these tools can be effectively employed during the early design stages of the PV inverter for optimal power electronic component selection.

Thus, it can be concluded that software tools are one of the key enablers of the DfR methodology, and they can help to identify critical weaknesses in the design of PV inverters early on in the product life cycle. The fast and user-friendly approach to the simulation and analysis of complex engineering problems make simulation tools widely used among design engineers. However, it should be noted that there is always a balance between computational-efficiency and simulation accuracy, which the designer should keep in mind when performing a given simulation and, inherently, during the overall design of the PV inverter. Moreover, the uncertainties introduced during the modeling process and the underlying assumptions of a given model or simulation (e.g., solver, sampling rate) should be clear and well-understood before the simulation results are interpreted and product design decisions or actions are taken.

3.5   Reliability testing and robustness validation

During the early development stages of a product, reliability analysis tools such as FMEA or DfR are used to ensure that the product design meets specific lifetime and safety criteria. Typically, after several design iterations, and only after the reliability requirements have been met, the product can move forward toward more mature product life cycle stages. Thus, based on the assumptions made during the design stage, a product prototype can be built. The main purpose of the prototype is not only to verify the correct functionality of the product but also to validate its reliability and robustness. Consequently, qualitative testing methods are usually employed to determine how robust the product design is, whereas quantitative testing methods are used to find the reliability performance of the product. Moreover, qualification (or design verification) testing is performed to ensure that an application-dependent relevant set of specific requirements (e.g., typically based on international standards) is met. A brief description of these three testing methods is given in Sections 3.5.1–3.5.3.

3.51   Qualitative test methods

Qualitative accelerated test methods are used to find the weakest points of an item of interest. By identifying these weaknesses, corrective measures can be taken, and the robustness of the product can be improved. Two of the most commonly used qualitative test methods are HALT and Highly Accelerated Stress Screening (HASS).

- **Highly Accelerated Life Testing**

  ○ HALT is an iterative testing process, which can improve the quality and robustness of the product, by identifying and removing the main weak-points in the design [31]. These tests are typically performed during more mature development stages, when a functional product-level, system-level, or even component-level prototype is already available. Through HALT, the operating and destructive limits of the product can be found, and, thus,

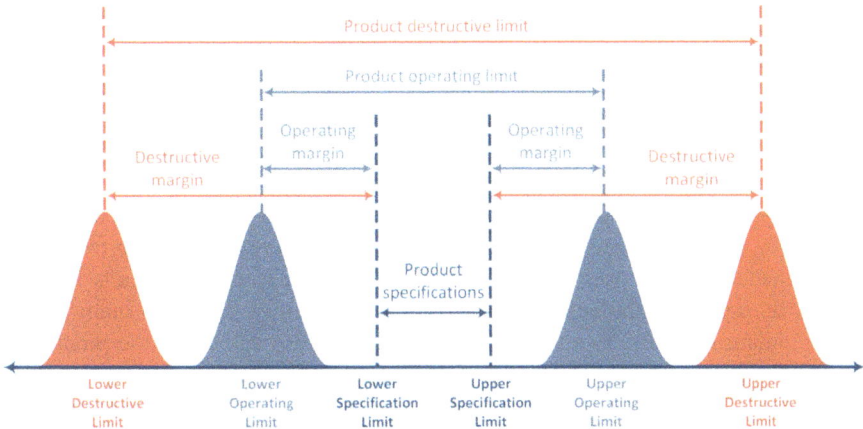

Figure 3.9   Samples Operating and Destructive Limits Outcomes of Halt for a Specific Stress Factor (E.g., Vibration or Shock and Thermal Cycling)

the engineer can ensure that enough design margin is available to account for possible variations (e.g., degradation, changes in environmental conditions) and that the product requirements are met. The main stress factors that are usually applied in HALT are temperature, thermal cycling, vibration or shock, voltage, and a combination of vibration or shock and thermal cycling. Multiple tests are performed until the margins between the operating and destructive limits are within a reasonable range, according to product and application requirements. Sample operating and destructive limits resulting from HALT are shown in Figure 3.9. Despite having the limitation of not providing any information on the product reliability, HALT is still one of the most widely used testing procedures due to its large saving in terms of warranty and service costs and useful insight regarding the strength and operational margins of the product. A detailed guideline on the applicability and implementation of HALT can be found in the IEC 62506 standard [32].

- **Highly Accelerated Stress Screening**

  ○ The operating and destructive limits of the product, found during HALT, can be further used in HASS tests to find product defects, which are introduced during the manufacturing process [31]. HASS testing consists of short exposures of the product at fast thermal cycles and vibration stress. The temperature and vibration stress level ranges are typically chosen outside of the product specification, and, thus, it is critical that HALT is performed prior to HASS. HASS testing implies that 100 percent of the production samples are screened, and, thus, it is most effectively applied for slow production rates or for the production of mission-critical products. Otherwise, for mature large-scale productions, an audit-based screening (e.g., only a given number of samples are screened) can be performed, and Highly Accelerated Stress Screening Audit (HASA) replaces HASS. Similar to HALT, neither HASS nor HASA can provide any information about the product reliability. Additionally, both have the limitation of increased risk of provoking new FMs and/or inducing "too much" degradation during the outside product-specification testing. However, HASS provides the benefit of being a quick testing method, which can detect faults induced during the manufacturing process, and, inherently, results in many of the infant mortality cases being screened out before delivering the product to the customer.

3.5.2   Quantitative test methods

Quantitative accelerated test methods are mainly employed to measure the reliability performance of a given product in a much shorter time than its expected lifetime. This can be achieved through accelerated tests, which consist of applying increased stress levels and/or repeated stress events to a product, with the purpose of quantifying the correlation between the test stress level and the exposure time. This can, in turn, provide the basis for developing an analytical lifetime model, which can be used to predict the reliability of a given product at a given use-case condition. However, prior to performing the accelerated tests, a clear understanding of the failure mechanisms, stressors, and environmental conditions must be ensured, so that they can accurately be reproduced within the accelerated test. Some of the most common accelerated test methods are the classic ALT and Calibrated Accelerated Life Testing (CALT). A brief description of the two is given below:

- Accelerated Life Testing

    Accelerated life testing is used to determine the impact that a given stressor has on the time-to-failure (or cycles-to-failure) of a product. Thus, considering a specific failure mechanism, the wear-out failure of a product can be addressed by calculating its lifetime under a given use-case condition, with the help of acceleration models and acceleration factors (AFs). By determining the time-to-failure (T_H) at an accelerated stress level (S_H), the time-to-failure (T_L) of the product under a lower accelerated stress level (S_L) can be calculated based on the AF [33].

$$T_L = AF \cdot T_H \tag{3.2}$$

Depending on the acceleration model used to describe a particular failure mechanism, the AF can be determined as the ratio between two different stress levels. Otherwise, in cases where the AF is unknown, acceleration models can be employed

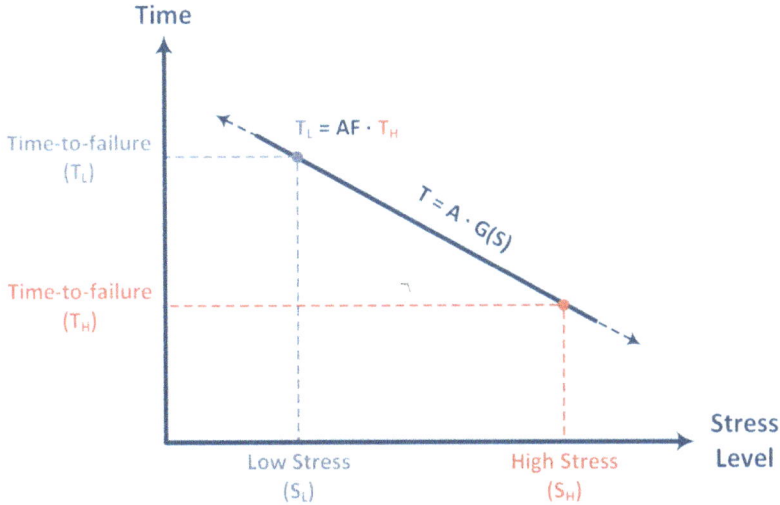

Figure 3.10   Basic principle of accelerated life testing (based on [33])

for calculating the time-to-failure at any given stress level, and it can be defined as the product between a constant model parameter (A) and the function of the stress level (G(S)).

$$T = A \cdot G(S) \tag{3.3}$$

Similar to the AF, the function of the stress level is dependent on the model, which best describes the considered wear-our failure mechanism (e.g., inverse power law, Arrhenius model, Eyring model), whereas the constant model parameters can be obtained by curve fitting the time-to-failure of at least three different stress conditions. The basic principle of the ALT method is presented in Figure 3.10.

A detailed description on the applicability and implementation of the accelerated life testing procedure can be found in [34] and in the IEC 62506 standard [32]. Since accelerated life testing methods are most effective when applied at the component or subassembly level, for grid-connected PV inverters, the main focus is usually placed on the most fragile parts, such as the IGBT power modules or the capacitors. For IGBT power modules, AC and DC power cycling test setups (as the one shown in Figure 3.11) are typically employed to expose the component to various accelerated stress levels until failure, and thus a particular failure mechanism can be induced, and the AF and acceleration model can be determined. Despite providing valuable insights regarding the product lifetime and a deep understanding of its reliability performance at different stress levels, accelerated life testing requires extensive knowledge of the product target failure mechanism, real-life stress conditions, etc. Additionally, since each failure mechanism needs to be dealt

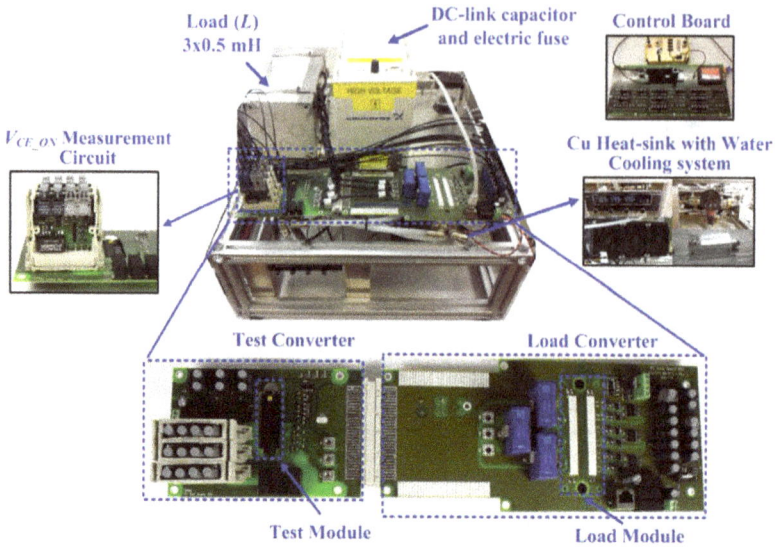

Figure 3.11   Advanced AC Accelerated Power Cycling Test Setup For IGBT Power Modules [35]

with separately, a large number of samples and stress conditions need to be tested until failure for the acceleration model and AF to have statistical significance. Thus, long testing time (e.g., up to a few months—depending on the selected stress level) can be expected.

- Calibrated Accelerated Life Testing

    For cases when the available testing time is limited, and the classic ALT cannot be effectively implemented, the CALT method can be employed. Similar to ALT, CALT is used to quantify the reliability performance of a product in terms of its expected lifetime for a given failure mechanism but in a much shorter testing time span. Prior to CALT, HALT needs to be performed to identify the overstress level margin of the product, thus identifying the foolish stress level. The procedure commences with the testing of two samples at a stress level, which is 10–15 percent lower than the foolish stress level. Afterward, a second test is performed, during which another two samples are tested at a stress level 10–15 percent lower than the foolish stress level. Based on the outcomes of the initial two tests, the third stress level can be determined by means of extrapolation. Considering the available testing time, the suitable stress level is selected and two or more samples can be tested to failure during the third and final testing phase. Finally, based on the testing results, the

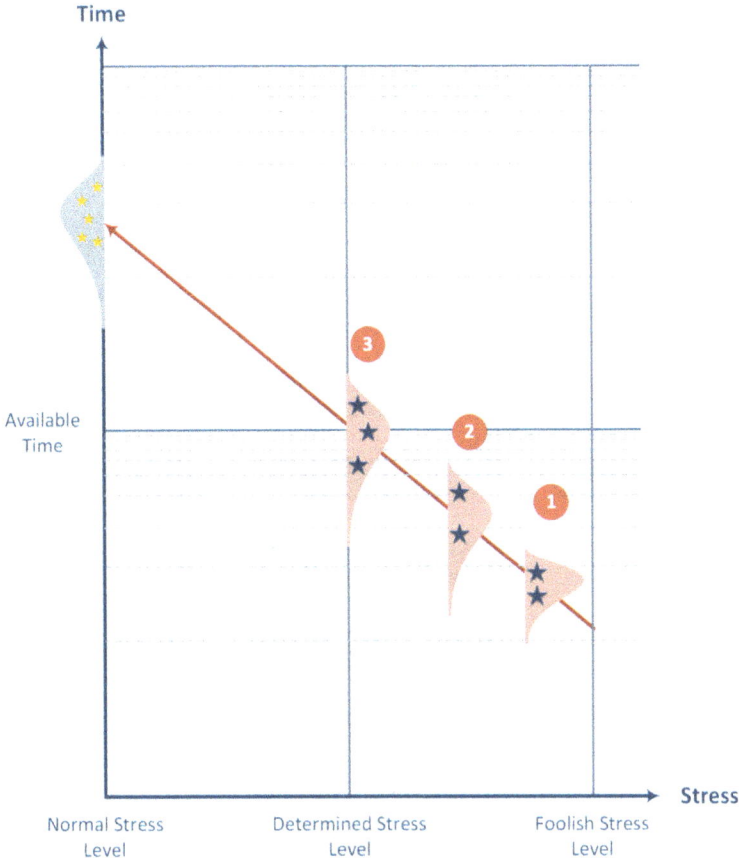

Figure 3.12   Basic principle of CALT procedure (based on [36])

acceleration model can be determined and the lifetime of the product at any given normal stress level (e.g., typical real-life conditions) can be estimated. The overall principle of the CALT method is presented in Figure 3.12, and a detailed description of its assumptions and implementation is given in [36].

3.5.3   Qualification testing

Qualification (or design verification) testing is meant to establish if a product can fulfill a given set of specification requirements (e.g., functionality, quality, safety) and if it meets the relevant international standards for a given application. This is typically achieved by exposing the product to a series of qualification tests, which, for grid-connected PV inverters, can be performed either at subcomponent level (e.g., IGBT power module, capacitor) or at system level (e.g., PV inverter).

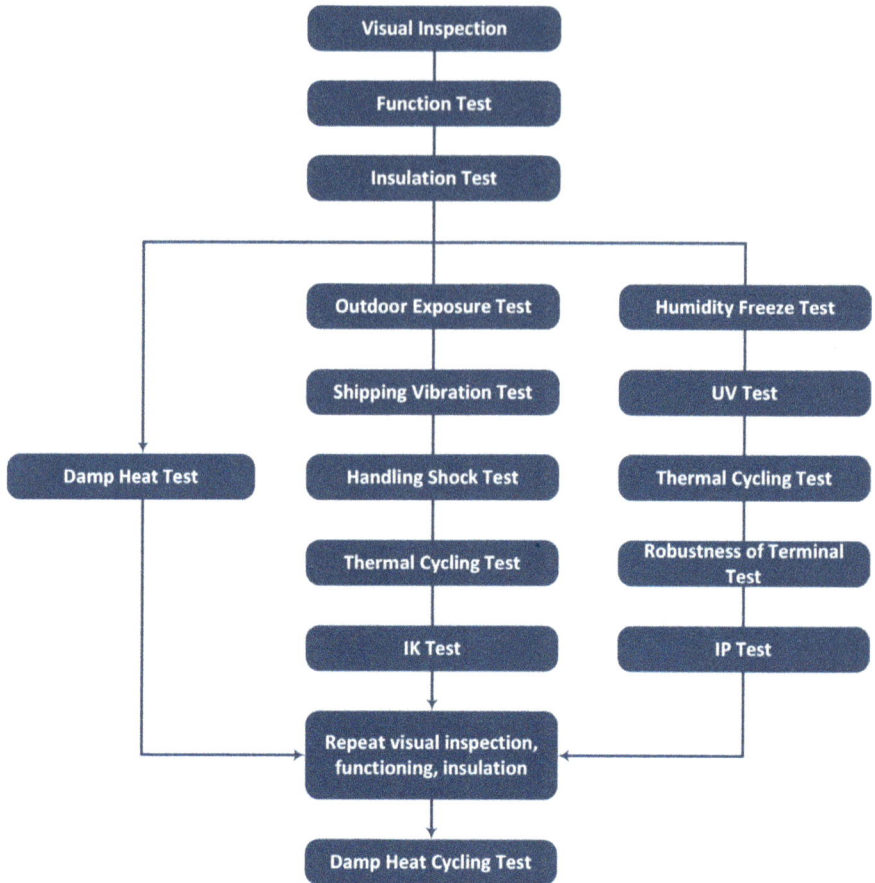

Figure 3.13   Qualification testing procedure for PV inverter, as defined in IEC 62093 [43]

Although qualification standards for IGBT power modules or capacitors rated for PV applications are still in their infancy, generic qualification testing standards (e.g., IEC 60747 [37], IEC 60749 [38], IEC 60068 [39], IEC 60384 [40], JESD 22 [41]) can be considered for reference. Power modules need to undergo a series of tests including power cycling, temperature cycling, high-temperature reverse bias, high-temperature high-humidity reverse bias, high-temperature and low-temperature storage test, as well as mechanical shock, to ensure a certain level of quality. Similarly, the capacitors need to pass a series of environmental and exposure qualification tests (e.g., thermal shock, damp heat, high temperature, vibration, charge, and discharge) to be considered as market-ready qualified [42].

On the other hand, for PV inverter systems, the qualification test standards are in a more mature stage of their development. A general overview of the qualification tests, which a PV inverter must be exposed to, is presented in Figure 3.13, as described in the IEC 62093 standard [43]. The PV inverter must undergo some basic visual inspection, functionality tests, and insulation tests, after being subject to each of the environmental qualification tests, thus, assuring that certain quality requirements are met, without sacrificing the intended operation of the inverter. Additional industrial and standardized qualification stress tests for PV inverters are summarized in [7]. Despite not proving any information regarding the reliability or robustness of the PV inverter, qualification tests are important tools that can be used to guarantee that certain levels of safety and quality are achieved for various use-cases and environmental conditions.

3.6. Summary

With the rapid integration and installation of more and more PV systems, the reliability of the PV inverter plays a crucial role in the safety, availability, and LCOE of the overall system. To deal with the ever-growing and stringent reliability requirements that the PV inverter must fulfill, a particular emphasis is placed on the accurate and effective reliability analysis of the inverter during the various stages of its development process. Thus, PV inverter manufacturers, integrators, and users have a wide array of reliability analysis methods and tools at their disposal.

During the early design and development stages, tools such as FMEA can be employed to identify the various FMs that occur at the different product levels of the PV inverter and to get a better understanding of their causes or effects and of their overall impact on the inverter's reliability and intended functionality. Additionally, enabled by various software simulation tools, the DfR methodology can be used to assess the reliability performance of the inverter under realistic environmental and operating conditions, thus, allowing for quick design weakness detection and for the reliability-oriented optimization of the inverter.

Further down the development process of the PV inverter, different testing methodologies can be applied to assess and improve its reliability and robustness. Qualitative testing methods, such as HALT or HASS, can be performed during the "Prototype" and "Manufacturing" life cycle stages to determine its operating and destructive limits and, inherently, assess the robustness margins of the PV inverter operation. Qualitative testing methods, such as ALT or CALT, can then be employed and used to quantify the reliability performance of the PV inverter under use-case conditions, by means of acceleration models and AFs. Before its market release, the inverter and its critical components must be subject to various qualification tests to assure that specific application-dependent requirements or specific international standards are fulfilled.

Finally, it can be concluded that there is not one specific tool or method that can address the reliability challenges faced by the design engineers of PV inverters. Rather, a combination of the above-mentioned tools should be used to get an accurate overview of the reliability and robustness performance of the PV inverter and, consequently, develop a sustainable reliability growth strategy throughout its entire development process. Due to the continuous development of more efficient testing methods, new standards, and the rapid progress of simulation software in terms of computational-efficiency and capabilities, the reliability assessment and allocation strategy for PV inverters should remain subject to continuous improvement.

References

[1]   World energy outlook 2017. *International Energy Agency [online]*. 2017. Available from https://www.iea.org/ [Accessed Nov 2020].

[2]   Fu R., Feldman D., Margolis R., Woodhouse M., Ardani K. U.S. Solar Photovoltaic System Cost Benchmark: Q1 2017, National Renewable Energy Laboratory (NREL), Tech. Rep. No. NREL/TP-6A20-68925. 2017. Available from https://www.nrel.gov/docs/fy17osti/68925.pdf [Accessed Nov 2020].

[3]   Wang H., Liserre M., Blaabjerg F., *et al.* 'Transitioning to physics-of-failure as a reliability driver in power electronics'. *IEEE Journal of Emerging and Selected Topics in Power Electronics*. 2014;**2**(1):97–114.

[4]   Wang H., Blaabjerg F. 'Power electronics reliability: state of the art and outlook'. *IEEE Journal of Emerging and Selected Topics in Power Electronics*. 2020:1.

[5]   Golnas A. 'PV system reliability: an operator's perspective'. *IEEE Journal of Photovoltaics*. 2013;**3**(1):416–21.

[6]   Hibberd B. *PV reliability and performance: a project developers experience.* [online] Proceedings of NREL Photovoltaic Module Reliability Workshop 2011, NREL/TP-5200-60170. Available from https://www.nrel.gov/docs/fy-14osti/60170.pdf [Accessed Nov 2020].

[7]   Hacke P., Lokanath S., Williams P., *et al.* 'A status review of photovoltaic power conversion equipment reliability, safety, and quality assurance protocols'. *Renewable and Sustainable Energy Reviews*. 2018;**82**(1):1097–112.

[8]   Moore L.M., Post H.N. 'Five years of operating experience at a large, utility-scale photovoltaic generating plant'. *Progress in Photovoltaics: Research and Applications*. 2008;**16**(3):249–59.

[9]   Lokanath S. *Central inverter cost of ownership & event analysis.* [online] Proceedings of the NREL/SNL/BNL PV Reliability Workshops. 2017. Available from https://www.nrel.gov/docs/fy17osti/68942.pdf [Accessed Nov 2020].

[10] 'Failure mode and effect analysis (FMEA and FMECA)'. *Standard IEC 60812*. 2018.

[11] Carlson C.S. in Hoboken N.J. (ed.) *Effective FMEAs: Achieving Safe, Reliable, and Economical Products and Processes Using Failure Mode and Effect Analysis*. USA: Wiley; 2012.

[12] AIAG. *Potential failure mode and effect analysis (FMEA): reference manual (4th edition)* [online]. 2008. Available from http://parsetraining.com/wp-content/uploads/2018/11/FMEA_Fourth-Edition.pdf [Accessed Nov 2020].

[13] Ma K., Liserre M., Blaabjerg F., Kerekes T. 'Thermal loading and lifetime estimation for power device considering mission profiles in wind power converter'. *IEEE Transactions on Power Electronics*. 2015;**30**(2):590–602.

[14] Burgos R., Chen G., Wang F., Boroyevich D., Odendaal W.G., Van Wyk J.D. 'Reliability-oriented design of three-phase power converters for aircraft applications'. *IEEE Transactions on Aerospace and Electronic Systems*. 2012;**48**(2):1249–63.

[15] Soldati A., Pietrini G., Dalboni M., Concari C. 'Electric-vehicle power converters model-based design-for-reliability'. *CPSS Transactions on Power Electronics and Applications*. 2018;**3**(2):102–10.

[16] Sangwongwanich A., Yang Y., Sera D., Blaabjerg F. 'Lifetime evaluation of grid-connected PV inverters considering panel degradation rates and installation sites'. *IEEE Transactions on Power Electronics*. 2018;**33**(2):1225–36.

[17] Vernica I., Wang H., Blaabjerg F. 'Uncertainty analysis of capacitor reliability prediction due to uneven thermal loading in photovoltaic applications'. *Microelectronics Reliability*. 2018;**88-90**(1):1036–41.

[18] Shen Y., Chub A., Wang H., Vinnikov D., Liivik E., Blaabjerg F. 'Wear-out failure analysis of an impedance-source PV microinverter based on system-level electrothermal modeling'. *IEEE Transactions on Industrial Electronics*. 2019;**66**(5):3914–27.

[19] ANSYS®. Available from www.ansys.com/products [Accessed Nov 2020].

[20] COMSOL®. Available from www.comsol.com/products [Accessed Nov 2020].

[21] MATLAB Simulink®. Available from www.mathworks.com/products/simulink [Accessed Nov 2020].

[22] PLECS®. Available from www.plexim.com/products/plecs [Accessed Nov 2020].

[23] SaberRD®. Available from www.synopsis.com/verification/virtual-prototyping/saber/saber-rd [Accessed Nov 2020].

[24] Reliasoft®. Available from www.reliasoft.com/ [Accessed Nov 2020].

[25] Evans T.M., Le Q., Mukherjee S., *et al.* 'PowerSynth: a power module layout generation tool'. *IEEE Transactions on Power Electronics*. 2019;**34**(6):5063–78.

[26] Simulation Assisted Reliability Assessment (SARA®). Available from calce. umd.edu/calce-sara-software [Accessed Nov 2020].

[27] Vernica I., Wang H., Blaabjerg F. 'A mission-profile-based tool for the reliability evaluation of power semiconductor devices in hybrid electric vehicles'. *Proceedings of IEEE International Symposium on Power Semiconductor Devices and ICs*; 2020. pp. 380–3.

[28] Vernica I., Wang H., Blaabjerg F. 'Design for reliability and robustness tool platform for power electronic systems – study case on motor drive applications'. *Proceedings of IEEE Applied Power Electronics Conference*; 2018. pp. 1799–806.

[29] SemiSel®. Available from semisel.semikron.com/#/home [Accessed Nov 2020].

[30] IPOSIM®. Available from www.infineon.com/cms/en/tools/landing/iposim. html [Accessed Nov 2020].

[31] Hobbs G.K. *Accelerated Reliability Engineering: HALT and HASS*. New York, USA: Wiley; 2000.

[32] 'Methods for product accelerated testing'. *Standard IEC 62506*. 2013.

[33] Otto S., Schmidt K.A., Johansen J. *Practically applicable reliability tools – A guide with practical cases*. SPM-181; 2013. Available from https://uk-spm. madebydelta.com/publications/reports/ [Accessed Nov. 2020].

[34] Kentved A.B. *Acceleration factors and accelerated life testing*. SPM-179; 2011. Available from https://uk-spm.madebydelta.com/publications/reports/ [Accessed Nov. 2020].

[35] Choi U.-M., Blaabjerg F., Jørgensen S. 'Power cycling test methods for reliability assessment of power device modules in respect to temperature stress'. *IEEE Transactions on Power Electronics*. 2018;**33**(3):2531–51.

[36] Edson L. *The GMW3172 Users Guide: Electrical Component Testing*; 2008.

[37] 'Semiconductor devices – Part 1: General'. *Standard IEC 60747*. 2006.

[38] 'Semiconductor devices – Mechanical and climatic test methods – Part 26: Electrostatic discharge (ESD) sensitivity testing – Human body model (HBM)'. *Standard IEC 60749*. 2018.

[39] 'Environmental testing – Part 2: Tests – ALL PARTS'. *Standard IEC 60068*. 2020.

[40] 'Fixed capacitors for use in electronic equipment – Part 16: Sectional specification – Fixed metallized polypropylene film dielectric DC capacitors'. *Standard IEC 60384*. 2019.

[41] 'Solid state device packaging standards'. *Standard JESD 22*. 2020.

[42] *Qualification of DC-link capacitors for automotive use: General requirements, test conditions and tests, ZVEI*; Frakfurt am Main, Germany; 2017.

[43] 'Balance-of-system components for photovoltaic systems – design qualification natural environments'. *Standard IEC 62093*. 2005.

Chapter 4

Grid-connected solar inverter system: a case study

V. S. Bharath Kurukuru[1]

Photovoltaic (PV) power systems have many vulnerable components whose lifecycle reliability is prone to relatively high risks. These risk indices generally deal with ambient environments, temperature, and power losses in the system. Meanwhile, the highly variable and uncontrollable solar irradiance, and PV system power input result in high electrical stresses for PV modules, which affect the operational lifecycles and power electronic converters. Consequently, these effects result in low system reliability when compared with the reliability of conventional generating plants. Besides, the high penetration of PV generation into the distribution networks leads to a detrimental effect causing reverse power flow and unacceptable rise in voltage at the distribution feeders. Generally, these overvoltages trigger the protection aspects of the inverter to disconnect from the grid, resulting in islanding operation of the PV system. This further contributes to abrupt voltage fluctuations, load shedding, and sudden changes in power flow. Therefore, the distribution networks associated with high-penetration PV generation have a risk of power outages and in turn increased maintenance costs. This necessitated the tools and methodologies to quantify reliability of the grid-connected PV systems. These reliability analysis tools and methods serve to evaluate the performance of PV systems and generate reliability indices, which further aid in achieving efficient design at the planning stage and determining the reduced cost and improved benefits at the operational phase. A systematic way to evaluate the reliability of grid-connected PV inverters is then presented in this chapter. The reliability analysis is carried out at the 2.2 MW grid-connected rooftop PV system installed in Jamia Millia Islamia (JMI), New Delhi, India, considering the variation of input power and failure rates of PV system components under ambient conditions. The reliability analysis is carried out for inverters at both string level and central level with the site data as a benchmark. A basic

[1]Advance Power Electronics Research Lab, Department of Electrical Engineering, Jamia Millia Islamia University, New Delhi, India

reliability indices computation method is adapted to realize the operation of both inverter systems for several risk metrics and quantify their impact on system operation.

4.1   Identifying the site for case study

PV systems have seen impressive growth around the world in the last decades. This growth is mainly due to the improving statistical average performance ratio of the new PV installations. However, the drawbacks due to probabilistic behavior and the unreliable nature of PV power demands the continuous improvement of the system. Besides, there are many vulnerable components in the PV system whose lifecycle reliability is highly affected by the operational conditions. Therefore, to improve the performance and increase the end of life of various components of PV systems, continuous monitoring and efficient reliability analysis methods need to be adapted. For facilitating the cost trade-off associated with PV systems, the reliability studies at the system and component level play a significant role. Generally, these studies are carried out at the laboratory level for various components of the PV system by subjecting them to accelerated testing scenarios. This process lacks various aspects when implemented with field operation of PV systems. Hence, meaningful statistical data of a PV plant operating under extreme climatic conditions with significant risks is the major aspect for performing the reliability analysis. Considering this, the PV installation in the northern part of India is identified for performing the case study.

4.1.1   Site details

The solar power industry in India is developing at a very fast pace with the lowest capital cost per MW. The Indian government is working toward an ambitious target of deploying 100 GW of solar PV capacity by 2022 under the national solar mission to combat climate change and avoid overdependence on fossil fuels for electricity generation [1]. By March 31, 2020, 37.6 GW had been installed [2], and rapid additional deployment is expected to occur over next years. The 100 GW target is part of the overall goal to achieve 175 GW of nonfossil fuel energy by 2022 and resonates well with India's COP-21 obligation in Paris to have 40 percent of its electricity generated by renewables by 2030. A recent announcement that the 175 GW goal may be increased to 227 GW further emphasizes India's commitment to renewables [3]. Out of the total installations, approximately 2.1 GW of PV systems correspond to rooftop solar power, which is the major aspect of this case study. Further, from the climatic point of view, the northern regions of India are considered to have extreme weather conditions that support the data required for the reliability analysis [4].

Given the above requirements, the reliability study of the grid-connected solar PV system in this chapter is focused on JMI central university located in the southeast part of Delhi, India. The university is in the coordinates $28°34'N\ 77°17'E$ with an overlap between semi-arid and monsoon-influenced humid subtropical with high variations between winter and summer temperatures [5]. The location

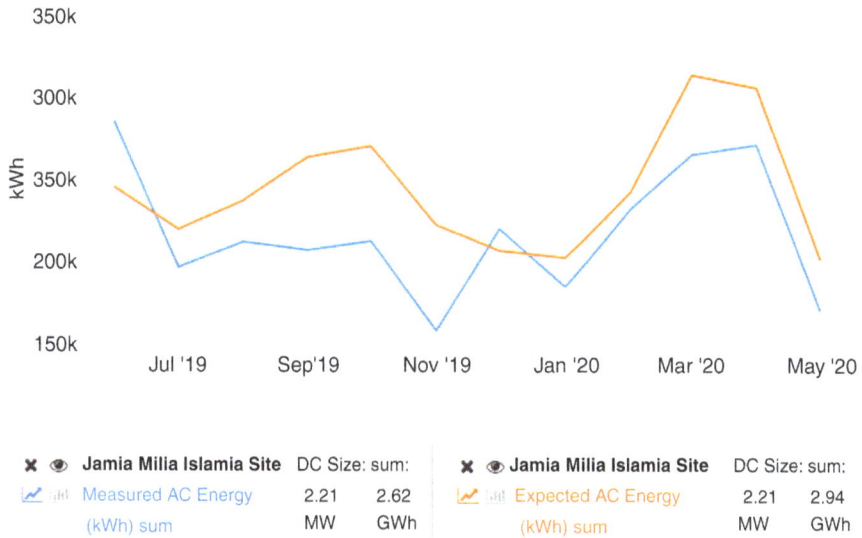

Figure 4.1   Measured and expected energy (kWh)

faces an extreme temperature ranging from −2.2°C to 48.4°C in four seasons [6]. The university has a 2.2 MW PV project developed by Sun Source Energy Pvt. Ltd. under renewable energy service company model. The PV installations are around 50 rooftops of various buildings in the campus such as the central library, colleges, departments, and hostels. The PV arrays on each rooftop are operated with string and central inverter-based systems. The performance of the total PV installations over the last 12 months (May 2019 to May 2020) is listed in Figures 4.1–4.8.

Figure 4.1 shows the measured and expected energy generation from the 2.2 MW plant. The expected energy corresponds to the predictions made based on the forecasted metrological data for the geographical location, and the measured data corresponds to the PV generation under the real operational conditions. From the graph, it is identified that the lowest expected peak is nearly around 200 kWp during January and the highest expected peak is nearly around 320 kWp during April. Further, the real-time measurements identified that the lowest peak is nearly around 150 kWp during November and the highest peak is around April. Moreover, it is identified that the sum of expected energy generation from May 2019 to May 2020 is 2.94 GWh, whereas the measured generation is 2.62 GWh for the same period.

Figure 4.2 shows the measured average power generated for every month from the 2.2 MW plant. It is identified that the maximum average generation occurs around February to April with 708 kW.

Figure 4.3 shows the measured maximum power generated for every month from the 2.2 MW plant. It is identified that the maximum average generation occurs during January with 18.5 MW.

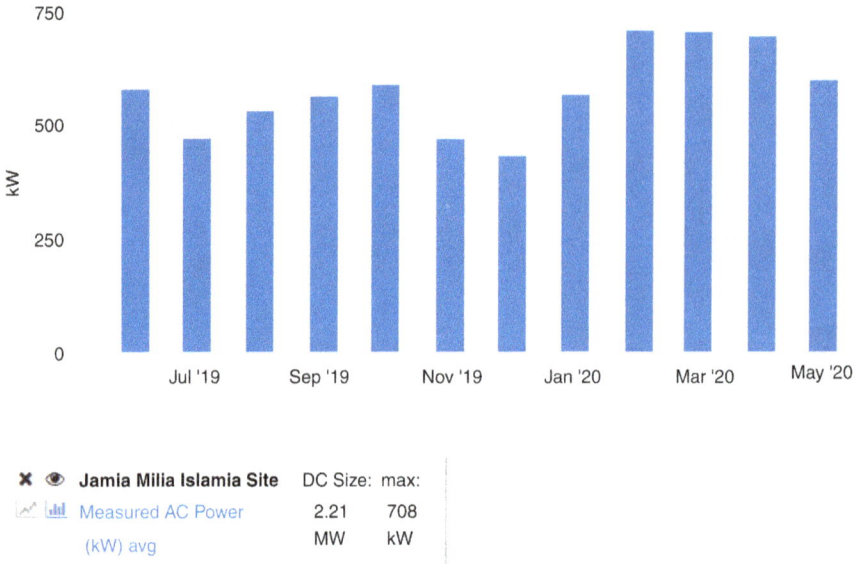

	Jamia Milia Islamia Site	DC Size:	max:
	Measured AC Power	2.21	708
	(kW) avg	MW	kW

Figure 4.2   Measured average power (kW)

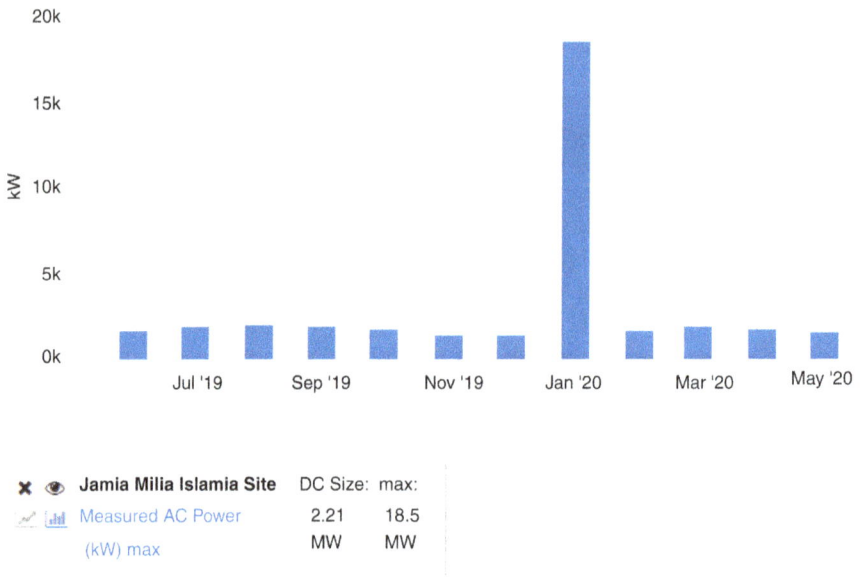

	Jamia Milia Islamia Site	DC Size:	max:
	Measured AC Power	2.21	18.5
	(kW) max	MW	MW

Figure 4.3   Measured maximum power (kW)

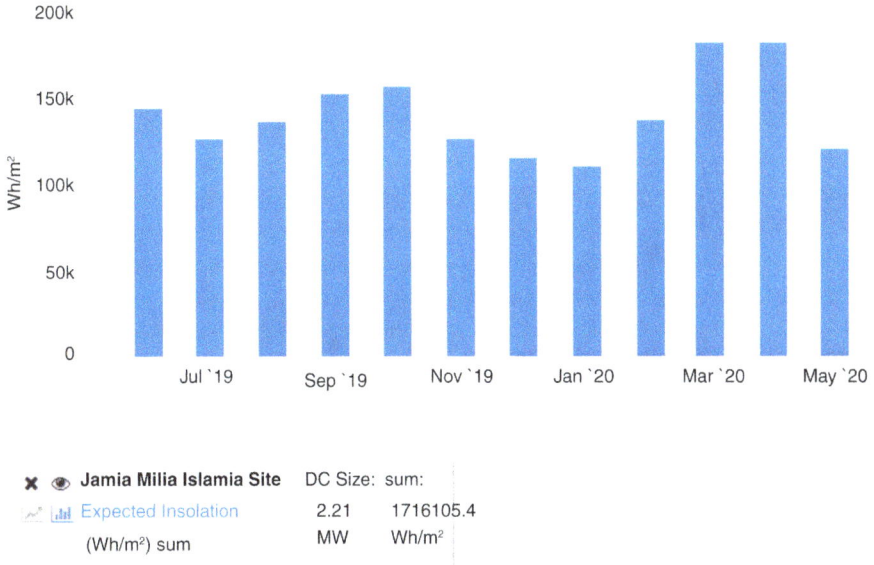

Figure 4.4   Sum of expected insolation (Wh/m²)

Figure 4.4 shows the sum of expected insolation for every month for the 2.2 MW plant. It is identified that the maximum expected solar insolation is around 170 000 Wh/m² during March and April. The sum of expected insolation is 1 716 105.4 Wh/m².

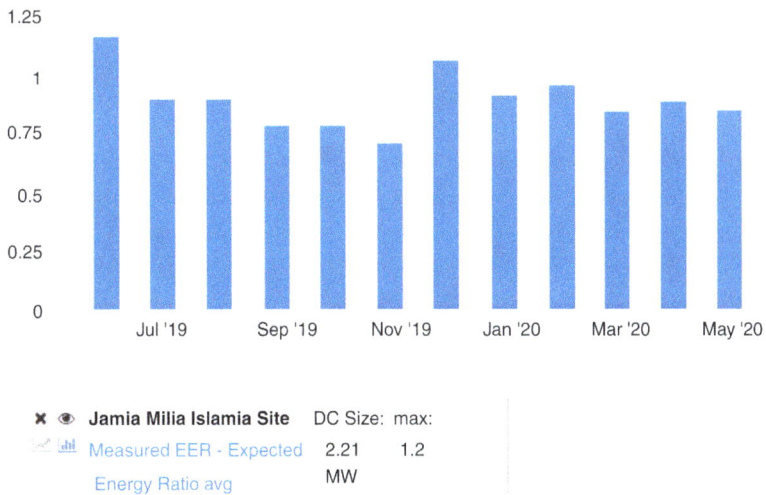

Figure 4.5   Average modeled expected energy ratio (EER)

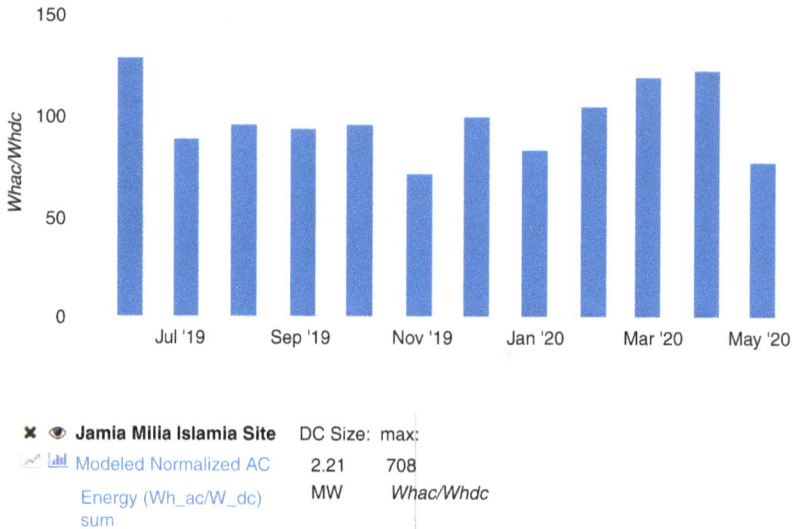

Figure 4.6   Sum of modeled normalized energy (Wh$_{ac}$/Wh$_{dc}$)

Figure 4.5 shows the EER for the monthly operation of the 2.2 MW plant. This performance ratio is measured as a ratio of expected output based on nameplate readings to the measured output for a given reporting period. From Figure 4.5, it is identified that the average EER from May 2019 to May 2020 is 1.2.

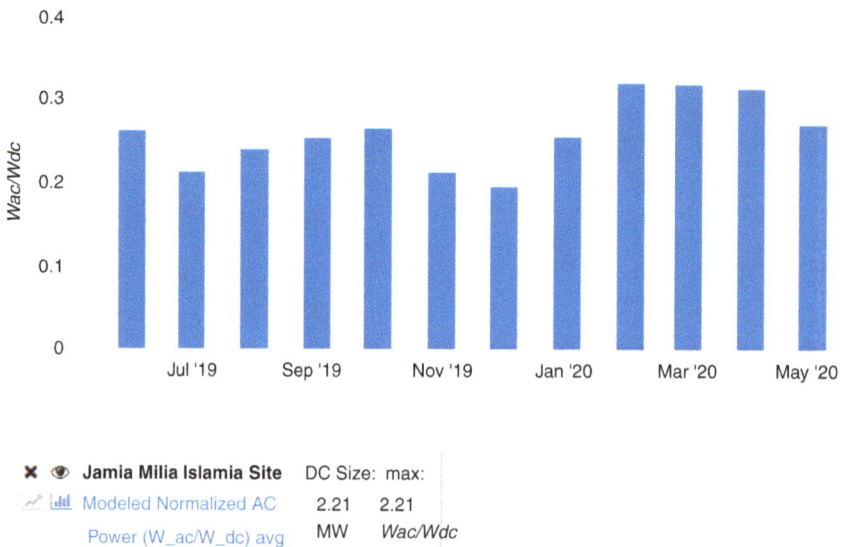

Figure 4.7   Average of modeled normalized power (W$_{ac}$/W$_{dc}$)

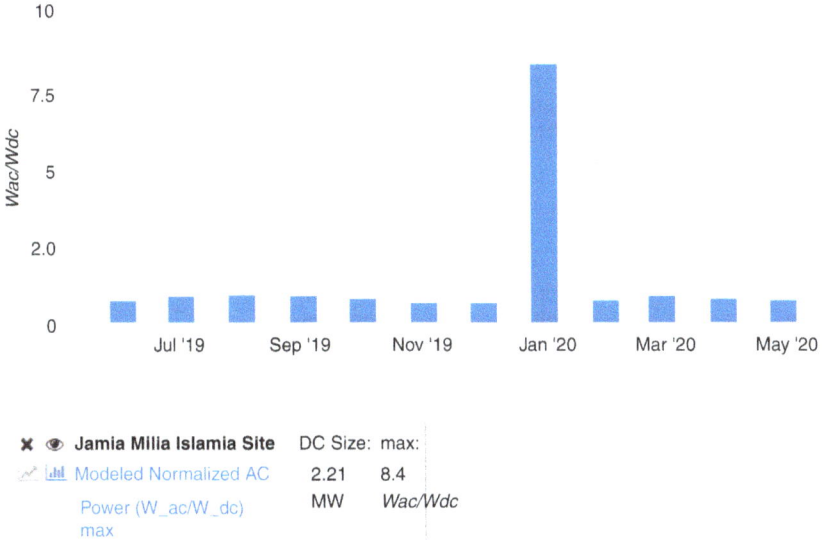

Figure 4.8   Maximum of modeled normalized power

Figure 4.6 shows the sum of normalized energy of the final yield per month for the 2.2 MW plant. This identifies the effect of temporal and local insulations that factor the reference yield obtained from the ratio of irradiated energy (kWh/m^2) to irradiance (kW/m^2), under standard test conditions (STC) for a given period [7]. For a given plant, the normalized energy is obtained as a ratio of plant output power to the nominal PV power generated under STC during a certain period. From the Figure 4.6, it is identified that the sum of modeled normalized energy is 1 187.6 Wh_{ac}/Wh_{dc}.

Figure 4.7 shows the sum of normalized power output per month for the 2.2 MW plant. This is characterized as a ratio of the total power output of PV generation to the nominal power generation capacity of the PV system under STC [7]. Generally, the average value of normalized power outputs varies from 0 to 1 and extends up to 2 depending on the cloud enhancements and PV installation area [7]. The average of normalized power is identified as 0.3 W_{ac}/W_{dc}. Further, the maximum normalized power shown in Figure 4.8 is identified as 8.4 W_{ac}/W_{dc}.

4.1.2   Components used in the system

The components used in this system are mainly PV modules and inverters of different ratings. For performing the reliability analysis of the components in the site, a base case of a PV system with 20 kW string and central inverters are considered. The details of the PV module and inverter are given as follows:

PV module: The chosen module is 310 W Sun-power type which has higher efficiency. Table 4.1 shows the characteristics of the Sun-power PV module.

Table 4.1   Characteristics of the Sun-power PV module

Parameter	Data
Manufacturer	Sun-power
Model	SPR-E19-310-COM
Efficiency	21.13%
Dimensions (module area)	Length: 1 559 mm
	Width: 1 046 mm
Specific parameters	Current at the maximum power: I_{mp} = 5.67 A
	Voltage at the maximum power: V_{mp} = 54.7 V
	Short circuit current: I_{sc} = 6.05 A
	Open circuit voltage: V_{oc} = 64.4 V

The reliability analysis is done for two different designs, a string inverter system of 20 kW and a central inverter system of 20 kW, where each system uses the same module rating of 310 W.

Inverter: Two different PV system configurations with 20 kW string inverter (SG15KTL-M/SG20KTL-M) and 20 kW central inverter (SG1250UD/SG1500UD) of Sungrow are chosen for conducting the reliability analysis. The details of both inverters are available in [8] and [9], respectively. The string inverter systems are mounted outdoor near the PV strings (either beneath a module or at a wall corner) and its temperature varies between 0 °C and 60 °C considering their direct exposure to heat emitted by PV panels and environmental conditions. Further, the central inverter is installed in an electrical room with additional cooling facilities and its temperature varies between 0 °C and 40 °C.

As the semiconductor modules are involved in the inverters, they are the most vulnerable components in the PV system [10]. These modules operate at high power levels and temperatures, which increase their failure risk and degrade the reliability of inverters [11, 12]. Hence, while performing reliability, sufficient data correspond-ing to the operation and failure rates of the PV inverter and PV modules is necessary. Sections 4.2 and 4.3 of the chapter discuss these aspects to identify the reliability indices and perform the reliability analysis.

4.2   Data collected for reliability study

The site identified for the case study achieves grid integration of PV systems by uti-lizing string and centralized inverter-based structures. As discussed in Section 4.1.2, a 20 kW string inverter and a 20 kW central inverter are considered for conducting the reliability assessment. The corresponding schematics of both systems are shown in Figures 4.9 and 4.10.

To distinguish, the string inverter system has each string connected to its inverter, whereas the central system has all the strings of an array connected to a single inverter. For the central system with total capacity equal to the capacity

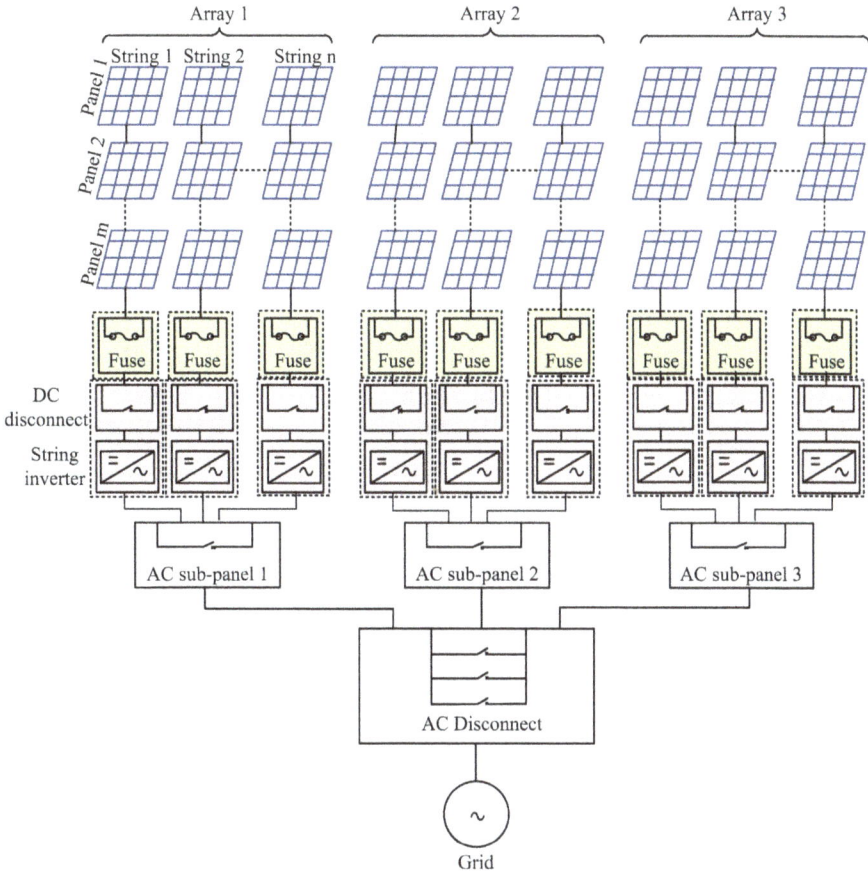

Figure 4.9 Schematic of a string inverter-based PV system

of an *n*-string system indicates that each string inverter capacity is $1/n$ of the central inverter. Further, from Figure 4.10, it is observed that, for a system with *n* PV strings, each string can generate $1/n$ of the power output of the total capacity. This indicates that failure in any string will not lead to system failure but will decrease the power output of the system. Considering this scenario, the reliability analysis is performed with an assumption that each PV string will have the same failure and repair rates. Further, the data required for the reliability analysis of a grid-connected solar inverter is obtained based on the failure rates of various components in the system.

4.2.1 Failure rate of power electronic switches

The failure rate of power electronics switches is calculated using empirical models discussed in [13–16]. Generally, these failure rates are determined by the thermal stress on the devices, e.g. insulated-gate bipolar transistors (IGBTs) and metal oxide semiconductor field-effect transistors (MOSFETs). This indicates that the failure

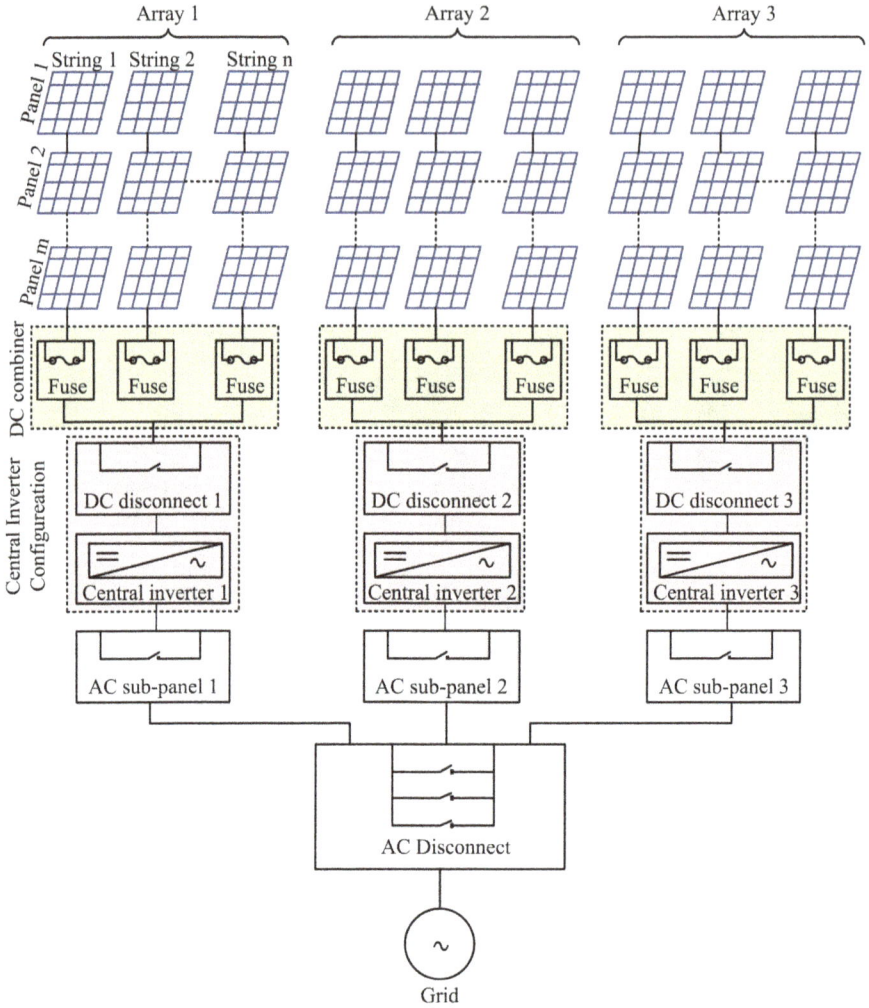

Figure 4.10   Schematic of a central inverter-based PV system

rate is a function of temperature or voltage, which is directly related to system input power levels and power loss. Besides, diodes are also associated with IGBTs and MOSFETs. Hence, their reliability also depends on system input power levels and power loss through voltage and temperature. The failure rate of diodes is calculated using empirical models discussed in [17].

4.2.1.1   Thermal model of IGBT and diode

Typically, IGBTs with anti-parallel diodes are widely used in PV inverters. The thermal models of an IGBT and a diode from junction to ambient and junction to case scenarios are given in Figures 4.11 and 4.12, respectively [18, 19]. For a known

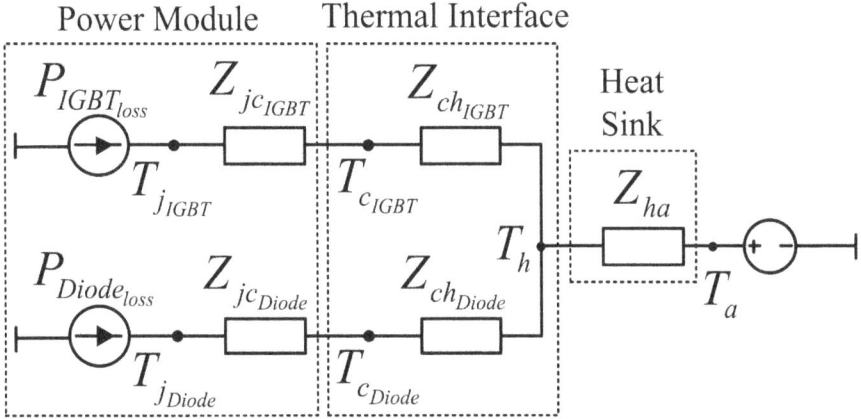

Figure 4.11 *Thermal model of IGBT and diode (single IGBT and diode with
T_j, T_c, T_h, and T_a corresponding to junction, case, heat sink,
and ambient temperatures, respectively, and Z_{jc}, Z_{ch}, and Z_{ha}
corresponding to thermal impedance junction to case, case to heat
sink, and heat sink to ambient, respectively)*

power loss in the system, the temperature variation of an IGBT and diode is estimated from the linear heat transfer equation discussed in [20] as

$$\Delta T_{IGBT} = R_{th1} P_{IGBT_{loss}} + R_{th2} P_{Diode_{loss}} \tag{4.1}$$

$$\Delta T_{Diode} = R_{th3} P_{IGBT_{loss}} + R_{th4} P_{Diode_{loss}} \tag{4.2}$$

where $P_{IGBT_{loss}}$ is the power dissipation in IGBT, $P_{Diode_{loss}}$ corresponds to power dissipation in the diode, R_{th1} and R_{th2} correspond to the thermal resistance of IGBT and
diode, respectively, and R_{th3} and R_{th4} correspond to thermal coupling coefficients
between IGBT and diode.

Further, the junction temperature of an IGBT or a diode is estimated as

$$T_j = T_c + \Delta T = T_a + R_{th} \left(P_{IGBT_{loss}} + P_{Diode_{loss}} + P_{add} \right) + \Delta T \tag{4.3}$$

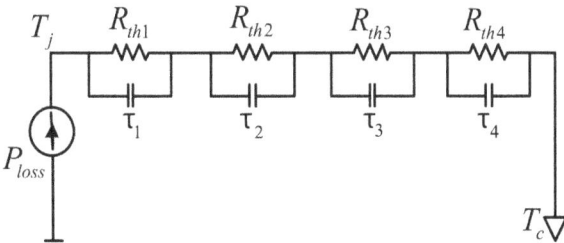

Figure 4.12 *Equivalent RC thermal network with R_{th} corresponding to resistance
and τ corresponding to capacitance [18]*

where T_c is the case temperature, T_a is the ambient temperature, R_{th} corresponds to thermal resistance from junction to the case including sink, and P_{add} is the additional power loss due to other mounted devices.

4.2.1.2   IGBT failure rate

The failure rate of an IGBT is estimated using an empirical model recommended by the Fides reliability guide 2009 [14] as

$$\lambda_{IGBT} = \left(\lambda_{0TH} \prod\nolimits_{Thermal} + \lambda_{0TCyCase} \prod\nolimits_{TCyCase} + \lambda_{0TCySJ} \prod\nolimits_{TCySJ} + \lambda_{0RH} \prod\nolimits_{RH} + \lambda_{0Mech} \prod\nolimits_{Mech}\right)$$

$$\prod\nolimits_{Induced} \prod\nolimits_{PM} \prod\nolimits_{Process} \tag{4.4}$$

where λ_{0TH} corresponds to the effect of thermal stress on the failure rate of an IGBT; $\lambda_{0TCyCase}$ corresponds to the effect of thermal cycling on the case; λ_{0TCySJ} corresponds to the effect of thermal cycling on the solder joint; λ_{0RH} and λ_{0Mech} correspond to humidity and mechanical overstress, respectively; $\prod_{thermal}$, $\prod_{TCyCase}$, $\prod_{TCySJ}$, $\prod_{RH}$, and $\prod_{Mech}$ correspond to physical overstress accelerating parameters of thermal, mechanical, and electrical origin; $\prod_{Induced}$ is the overstress caused by additional factors; $\prod_{PM}$ is the manufactured part quality; and $\prod_{Process}$ is the technical control over reliability and quality in a product life cycle.

For a known junction temperature, the temperature factor is calculated as

$$\prod\nolimits_{Thermal} = \prod\nolimits_{EI} . e^{11604 \times 0.7 \times \left[1/293 - 1/(T_j + 273)\right]} \tag{4.5}$$

where T_j corresponds to the IGBT junction temperature and

$$\prod\nolimits_{EI} = \begin{cases} \left(\dfrac{v_{applied}}{v_{r,IGBT}}\right)^{2.4} & if \left(\dfrac{v_{applied}}{v_{r,IGBT}}\right) > 0.3 \\ 0.056 & if \left(\dfrac{v_{applied}}{v_{r,IGBT}}\right) \leq 0.3 \end{cases} \tag{4.6}$$

in which $v_{applied}$ corresponds to the voltage applied across the IGBT, and $v_{r,IGBT}$ corresponds to the IGBT rated reverse voltage.

From (4.1) – (4.6), the failure rate of an IGBT is a function of voltage or temperature that corresponds to the power loss and system input power levels.

4.2.1.3   Diode failure rate

The failure rate of the diode in PV inverters is estimated using a standard reliability model discussed in [17] as

$$\lambda_D = \lambda_b \pi_T \pi_S \pi_C \pi_Q \pi_E \tag{4.7}$$

where λ_b represents the diode basic failure rate, π_T, π_S, π_C, π_Q, and π_E correspond to temperature, electrical stress, construction, quality, and environmental factors, respectively. For a known junction temperature T_j, the temperature factor is calculated as

$$\pi_T = e^{-3091\left(1/(T_j+273)-1/298\right)} \qquad (4.8)$$

The electrical stress factor [21, 22] is calculated as

$$\pi_S = \begin{cases} \left(\dfrac{v_{applied}}{v_{r,diode}}\right)^{2.45} & if\ 0.3 < \dfrac{v_{applied}}{v_{r,diode}} < 1 \\ 0.054 & if\ \dfrac{V_{applied}}{V_{r,\,diode}} \le 0.3 \end{cases} \qquad (4.9)$$

where $v_{applied}$ is the voltage applied across the diode and $v_{r,diode}$ corresponds to the diode-rated reverse voltage.

From the (4.7) – (4.9), the failure rate of a diode is similar to the failure rate of an IGBT and forms a function of voltage or temperature that corresponds to the power loss and system input power levels.

4.2.1.4 Capacitor failure rate

Another important factor that leads to the PV inverter failure is related to capacitors [23]. A comparative analysis in [24] has identified that the electrolytic capacitor dominates in an inverter failure. Further, the industrial representatives at US Department of Energy (DOE) workshop [25, 26] indicated that the quality of DC-bus capacitors is a critical problem affecting the reliability of the inverter. The conventional methods for predicting the reliability of capacitors [13] have identified that the capacitor failure rate is dependent on ripple current, DC voltage applied, and ambient conditions such as heat sinking, temperature, and airflow. Further, the inverters mounted outdoor are exposed to harsh ambient environments and may suffer high capacitor failure rates. This indicates that the capacitor failure rate can be determined by hotspot temperature, which is estimated by the base life at actual and maximum hotspot temperatures [27]. The generalized expression for computing the capacitor failure rate is given as

$$\lambda_C = \frac{1}{r_C} = \frac{1}{L_b \cdot 2^{(T_{max}-T_c)/10}} \qquad (4.10)$$

where r_C corresponds to the capacitor life expectancy, L_b corresponds to the base life at a maximum core temperature T_{max}, and T_C corresponds to the actual core temperature. Further, the lifetime of the capacitor depends on the ripple current flowing through it and can be estimated as a function of the hotspot temperature in (4.10). The current ripple rate for a central inverter-based system without storage components is approximated as

$$i_r\,(t) = \frac{V_0}{V_d} I_0 \cos\,(2\omega t - \varphi) \qquad (4.11)$$

in which V_0 is the grid voltage, V_d is the input DC voltage, I_0 is the RMS output current, and ω and φ correspond to the fundamental frequency and power factor, respectively. It is noted that the higher-order harmonics due to smaller amplitudes of on/off switching are neglected here. Further, the RMS value of the ripple current is calculated as

$$I_r = \frac{P_0}{\sqrt{2}V_d} \tag{4.12}$$

where P_0 denotes the inverter output power.

For a steady-state condition, the hotspot temperature of the capacitor is calculated as

$$T_c = T_a + \theta_c \left(I_r^2 R_s\right) \tag{4.13}$$

where T_a corresponds to the ambient temperature, θ_c represents the thermal resistance of the capacitor, and R_s corresponds to the equivalent series resistance (ESR) of the capacitor. Further, the power loss is obtained by substituting (4.13) into (4.10).

4.2.2   Reliability of inverter

Failure in any component of the PV inverter sometimes may lead to a complete outage indicating no parallel redundancy [11]. Hence, its reliability can be modeled as a series network with the reliability indices as

$$\lambda_I (P, V, T) = \lambda_C + \sum_i \left(\lambda_{Di} + \lambda_{Si}\right) \tag{4.14}$$

$$r_I (P, V, T) = \frac{1}{\lambda_I} \left[\lambda_C r_c + \sum_i \left(\lambda_{Di} r_{Di} + \lambda_{Si} r_{Si}\right)\right] \tag{4.15}$$

$$A_I (P, V, T) = \frac{1}{\lambda_I} \left[\frac{\frac{1}{r_1}}{\lambda_I + \frac{1}{r_1}}\right] \tag{4.16}$$

where λ_I represents the failure rate, r_I represents the repair time, and A_I represents the availability of PV inverter. Further, the subscripts C, D, and S correspond to the capacitor, diode, and IGBT, respectively, and i indicates the ith component. Besides, the availability of AC subpanel and DC disconnect is calculated from the failure and repair rate as

$$A_{DC} = \frac{\frac{1}{r_{DC}}}{\lambda_{DC} + \frac{1}{r_{DC}}} \tag{4.17}$$

$$A_{AC} = \frac{\frac{1}{r_{AC}}}{\lambda_{AC} + \frac{1}{r_{AC}}} \tag{4.18}$$

As the failure rate probability of the three-phase AC disconnect is very low, it can be assumed to be perfectly reliable and can be easily modeled if the failure data is available. Further, the reliability parameters for central and string configurations of a PV system are given in Tables 4.2 and 4.3, respectively [28].

Table 4.2   Base case reliability analysis parameters (central inverter system)

IGBT and diode

T_a=25 °C	μ=0.8	$cos\varphi$=0.95	V_t=480 V	f=20 kHz
P_{add}=11 W	θ_a=0.11 °C/W	V_{lo}=0.9654 V	a1=0.1642	b_j=0.6468
k=1.6783	z=0.0181	h=0.0040	r=1.3444	$V_{r,IGBT}$=480 V
θ_{11}=0.640 °C/W	θ_{12}=0.250 °C/W	θ_{21}=0.300 °C/W	θ_{22}=0.830 °C/W	r_{Di}=20 days
λ_{OTH}=0.3021	$\Pi_{induced}$=2.0	Π_{PM}=1.7	$\Pi_{Process}$=4.0	r_{Si}=20 days
k_{goff}=1.0	k_{gon}=1.5	$V_{r,diode}$=600 V	t_a=25.9 ns	t_b=54.1 ns
V_{2o}=0.711 V	a_2=0.136	b_2=0.395	I_{rr}=10 A	λ_b=0.005
π_E=6	π_c=1	π_Q=2.4		

Capacitor

L_b=20 000 hours	T_{max}=95 °C	R_s=0.02 ohms θ_c=15.6 °C/W	r_c=10 days

PV array

λ_{Pi}=1.1416	r_{Pi}=48	λ_F=5.7078	r_F=10	d=0.5% a=26.99	β=5.83

DC disconnect and AC subpanel

λ_{DC}=0.05	r_{DC}=16	λ_{AC}=0.01	r_{AC}=10

It should be noted that the unit for failure rates is 1/(106 hrs) and repair time is hrs. Further, these parameters are adapted with reliability evaluation techniques and the IEEE reliability standards [29] to assess the reliability of the PV plant. There are many techniques such as Monte Carlo simulations [30], Markov chain method [31], reliability block diagrams [32], fault tree analysis [33], and state enumeration method (SEM) [34] available in the literature. In this chapter, the SEM [35, 36] is used to perform the reliability analysis on the PV system. This method adapts the impact of voltage levels, input power levels, and power losses on the failure rate of different components in a PV system. It works on an underlying assumption that

Table 4.3   Base case reliability analysis parameters (string inverter)

IGBT and diode

T_a=25 °C	μ=0.8	$cos\varphi$=0.95	V_t=480 V	f=20 kHz
P_{add}=11 W	θ_a=0.11 °C/W	V_{lo}=0.9654 V	a_j=0.1642	b_j=0.6468
k=1.6783	z=0.0181	h=0.0040	r=1.3444	$V_{r,IGBT}$=480 V
θ_{11}=0.640 °C/W	θ_{12}=0.250 °C/W	θ_{21}=0.300 °C/W	θ_{22}=0.830 °C/W	r_{Di}=20 days
λ_{OTH}=0.3021	$\Pi_{Induced}$=2.0	Π_{PM}=1.7	$\Pi_{Process}$=4.0	r_{Si}=20 days
k_{goff}=1.0	k_{gon}=1.5	$V_{r,diode}$=600 V	t_a=25.9 ns	t_b=54.1 ns
V_{2o}=0.711 V	a_2=0.136	b_2=0.395	I_{rr}=10 A	λ_b=0.005
π_E=6	Π_c=1	Π_Q=2.4		

Capacitor

L_b=20 000 hours	T_{max}=95 °C	R_s=0.2 ohms	θ_c=4.52 °C/W r_c=10 days

PV array

λ_{Pi}=1.1416	r_{Pi}=48	r_F=10	d=0.5%	β=5.83

DC disconnect and AC subpanel

λ_{DC}=0.05	λ_{AC}=0.01 r_{AC}=10

each PV system has three operating states, the normal or idle state, the working state, and the out of service state. Initially, the equivalent reliability parameters of multiple PV string in an array are identified and their reliability indices are determined using SEM. Generally, these indices are either energy-oriented or time-oriented. The methodology corresponding to states of PV array and the reliability indices for both string and central system are discussed in Section 4.3.

4.3   Reliability study for the identified site

In this section, the methodology for conducting a reliability study on a base case of the identified site is presented. The failure rates are discussed, and the data collected in the previous section are the inputs to carry out the reliability analysis of grid-connected PV systems.

Algorithm 4.1: Approach for PV system risk analysis

Step 1:   Identify the chronological data of PV system (solar insolation, ambient
            temperature, power output, DC voltage, AC voltage, and frequency)
Step 2:   Evaluate the discrete probabilistic distribution of power input
Step 3:   • Assess the energy losses along with switching and conduction losses at
              each input power level in semiconductors.
            • Assess the energy losses in capacitors using DC voltage and ripple current.
            • Predict hotspot temperatures by thermal models of the IGBT, diode, and
              capacitor.
            • Quantify risks in PV inverters
Step 4:   State enumeration-based PV array risk analysis
Step 5:   Calculate the PV system risk indices and perform sensitivity analysis
            (insolation, temperature, number of strings, component reliability
            parameters, etc.)

4.3.1   Risk modeling for components of PV system

As discussed earlier in Section 4.2, the grid-connected PV systems either deal with string/multi-string or centralized structures. The string/multi-string structure deals with an own inverter for each string or multiple strings in the PV array. Whereas, for a central system, all the strings are connected to a single inverter system that is feeding the grid. For a condition where an *n* string inverter system has the same capacity of a central inverter system, the capacity of the central inverter system is *n* times of each string inverter system. The systematic approach adopted for PV risk analysis is shown in Algorithm 4.1.

4.3.1.1   The periodic discrete probability distribution of input power

The varying energy losses in the components connected to PV systems due to the periodic input power resulted in temperature variations in a solar inverter. Hence, the input power levels play a significant role in determining the life

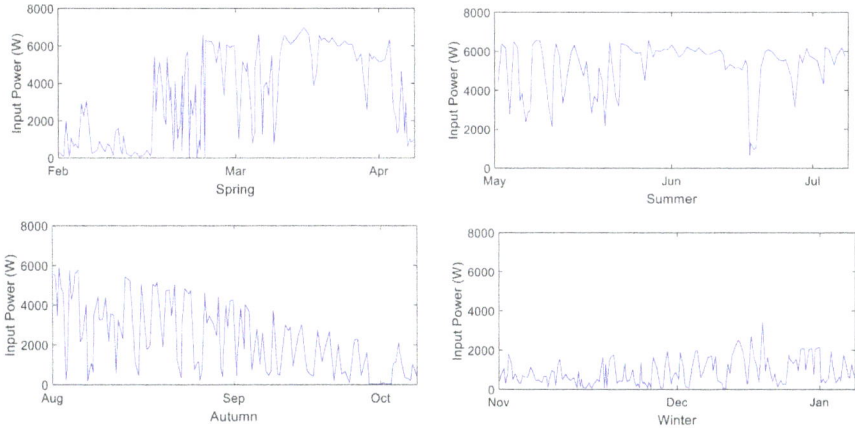

Figure 4.13   *Power curve data of different periods in chronological order of spring, summer, autumn, and winter*

cycle of the PV system and its components. For the site considered in this case study, the data loggers sample and record the operational quantities of the PV system for every 5–15 minutes. This helps in obtaining a linear and highly intermittent power curve with a broad set of data points. Further, to perform the periodic risk analysis, the power curve data is divided into four seasons as shown in Figure 4.13.

These input power curves are aggregated into a discrete probability distribution for quantifying their effect on PV system operation risks. The K-means clustering technique is used to cluster the data point into multiple power levels by eliminating the chronology. The detailed procedure is as follows: Initially, it is assumed to divide the annual power curve into K power levels. The term K is adjusted depending on the degree of precision needed for the reliability analysis. For the system in the case study, the satisfactory results are guaranteed by setting K around 10–15. Further, the K power level clustering of the yearly input power curve data with N data points is achieved by following Algorithm 4.2.

For a successful convergence, the mean μ_i corresponds to ith mean power level and the probability p_i is given by $p_i = \frac{N_{S_i}}{N}$, where N corresponds to the number of power curves.

The discrete probability distribution achieved by aggregating the periodic data of the input power curve in Figure 4.13 into 12 power levels is shown in Figure 4.14. Each power level in Figure 4.14 evaluates the availability of power electronic components, expected periodic energy output, and PV system risk indices weighted by the power level probability.

Algorithm 4.2: K power level clustering

Step 1: Initialize cluster $S = \{S_1, S_2, \ldots, S_k\}$ and randomly assign the data points to each cluster.

Step 2: Calculate mean for an initial cluster, where $i = 1, 2, \ldots, 3$ corresponds to the cluster S_i.

Step 3: Calculate the distance between each data point P_j ($j = 1, 2, \ldots, N$) and mean μ_i of ith cluster. $d_{ji} = |P_j - \mu_i|$

Step 4: The data points are assigned to the nearest cluster and the cluster means are recalculated using $\mu_i = \frac{1}{N_{S_i}} \Sigma p_j \epsilon S_i P_j$, where N_{si} corresponds to total data points in ith cluster.

Step 5: Iterate the process in steps 3 and 4 until every μ_i is unchanged between any two iterations.

4.3.2   *Reliability analysis of PV array*

4.3.2.1   Equivalent parameters for reliability of PV string

The PV string consists of series-connected PV modules with a fuse inside the DC combiner box. There are three repairable failure modes of PV modules [37–40]: (i) short circuit and (ii) open circuit of PV modules, and (iii) failure at a junction box

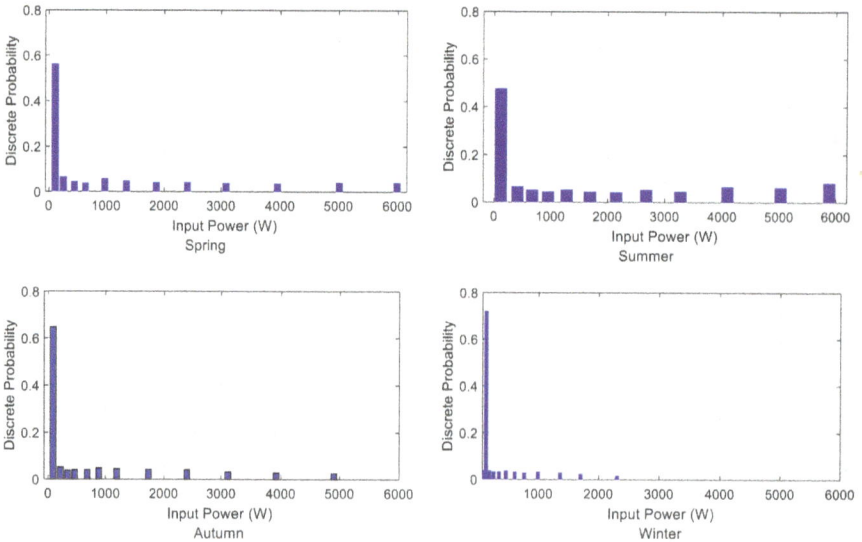

Figure 4.14   Discrete probability distribution of power curve data of different periods in chronological order of spring, summer, autumn, and winter

that results in an outage of the whole string. These faults can be characterized by their average failure rates and repair rates of the PV module. Further, because of shading effects, a PV module can be bypassed through diodes resulting in a lower output of the string. Hence, this phenomenon is not considered as an outage but instead represents a low input power level. Besides, the probability of the instantaneous bypass of multiple panels is low and can be negligible. Therefore, the equivalent parameters for the reliability of a PV string are given as

$$\lambda_S = \sum_{i=1}^{m} \lambda_{P,i} + \lambda_F \qquad (4.19)$$

$$r_S = \frac{1}{\lambda_S} \left(\sum_{i=1}^{m} \lambda_{P,i} r_{P,i} + \lambda_F r_F \right) \qquad (4.20)$$

where λ is the failure rate, S represents equivalent PV string, m corresponds to total number of modules in a string, P is an equivalent PV module, F indicates the fuse in DC combiner, r corresponds to repair time, and $\lambda_{P,i}$ and $r_{P,i}$ correspond to the failure and repair rate of the ith PV module, respectively.

4.3.2.2 State enumeration for PV array reliability analysis

The SEM is adapted to compute the PV array reliability parameters from the data of n PV strings. This method can be applied to both heterogeneous and homogenous PV strings. Generally, the availability, unavailability, and other operating states of PV strings in an array can be given as

$$\left(A_{S1} + U_{S1}\right)\left(A_{S2} + U_{S2}\right)\ldots\left(A_{Sn} + U_{Sn}\right) \qquad (4.21)$$

where A_{Si} is the availability and U_{Si} is the unavailability of the ith PV string and n corresponds to the number of PV strings in an array. Further, the availability of the ith string is calculated as

$$A_{Si} = \frac{\dfrac{1}{r_{Si}}}{\lambda_{Si} + \dfrac{1}{r_{Si}}} \qquad (4.22)$$

and the unavailability of the ith string is calculated as

$$U_{Si} = \frac{\lambda_{Si}}{\lambda_{Si} + \dfrac{1}{r_{Si}}} \qquad (4.23)$$

Considering the above aspects of the PV array, the probability of an enumerated state α is given as

$$p_A(\alpha) = \prod_{i=1}^{n_f} U_{Si} \prod_{i=1}^{n-n_f} A_{Si} \qquad (4.24)$$

with n_f representing the number of the failed PV strings and $n - n_f$ corresponds to healthy PV strings in state α.

For the condition where j PV strings fail, all the enumerated states are aggregated into the ith state of the PV array. This gives the probability of enumerated states as

$$P_{Aj} = \sum_{\alpha \in G(n_f = j)} P_A(\alpha) = \sum_{\alpha \in G(n_f = j)} \left(\prod_{i=1}^{n_f} U_{si} \prod_{i=1}^{n-n_f} A_{Si} \right) \quad j = 0, 1, \ldots, n \qquad (4.25)$$

where $G(n_f = j)$ corresponds to the enumerated states of j strings outage. Based on the above observations, state 1 corresponds to the outage of one string with $(n - 1)$ contingency, state 2 corresponds to the outage of two strings with $(n - 2)$ contingency, state j corresponds to the outage of j strings with $(n - j)$ contingency, and state n corresponds to the outage of all the PV strings. Additionally, the common causes of failures such as lightning, environmental effects, and mechanical issues or other electrical problems that are independent of n strings failure are represented as a downstate or an additional failure event in the enumeration process.

This discussion on state enumeration for PV array incorporates the impact of voltage levels, power inputs, failure rates, and power losses to provide a viable approach for the reliability analysis.

4.3.2.3   Effect of aging and degradation

The failures in PV panels and degradation of PV modules increase with an increase in operation time and advancing age. Hence, these factors play a significant role in risk analysis while dealing with the end-stage life of PV modules. Considering this, a linear model of PV panel degradation is developed as discussed in [17] and the performance degradation of the PV panel in terms of power for a lifetime is given as

$$p_i = p_0 \left[1 - (k - 1) d \right] \quad k = 1, 2, \ldots, L \qquad (4.26)$$

where p_0 corresponds to the PV module's initial power capacity, d represents the constant slope, and k corresponds to a specified year during the observed life cycle L.

Further, to assess the annual unavailability, an aging failure model is adapted as follows: for an aging failure, the probability function of failure density $f(t)$ with failure transition period t after surviving for T years is given by

$$P_{T,t} = \frac{\int_T^{T+t} f(t) \, dt}{\int_T^{\infty} f(t) \, dt} \qquad (4.27)$$

Further, the subsequent failure period t is divided into N subintervals with equal length Δx. This gives the failure probability at the ith interval as

$$P_i = \frac{\int_T^{T+i\Delta x} f(t) \, dt - \int_T^{T+(i-1)\Delta x} f(t) \, dt}{\int_T^{\infty} f(t) \, dt} \quad (i = 1, 2, \ldots, N) \qquad (4.28)$$

The average duration of unavailability for failures at the ith interval can be estimated as

$$UD_i = t - (2i - 1)\,\Delta x/2 \qquad (i = 1, 2, \ldots, N) \tag{4.29}$$

Further, the unavailability for a subsequent failure period t is given by

$$U_{T,t} = \sum_{i=1}^{N} P_i.UD_i/t \tag{4.30}$$

From (4.27) to (4.30), the total unavailability of repairable and non-repairable aging failures with the failure transition period t after surviving for T years is given as

$$U_S = U_{S,r} + U_{T,t} - U_{S,r}U_{T,t} \tag{4.31}$$

$$A_S = 1 - U_S \tag{4.32}$$

where U_S and A_S correspond to the total unavailability and availability of a PV string, respectively.

4.3.3   PV system risk indices

The PV risk indices quantify the performance of the PV system and are useful at the planning stage for design selection and at the operation stage for reduced costs and increased benefits. Conventionally, the outage duration and failure rate are widely adopted as PV system risk indices. Further, two new indices that are focused on energy and time are discussed for better performance quantification.

4.3.3.1   Equivalent parameters for reliability of PV string

The energy-based indices provide the annual statistics of the PV system energy yield along with the system uncertainties.

A. Ideal energy output

The ideal energy output (IEO) estimates the power output of a PV system by multiplying the clustered power levels of the perfectly reliable generation with their corresponding converter efficiency curves. This is mathematically expressed as

$$\text{IEO} = \sum_{l}\sum_{i=1}^{K} \mu_i \eta_i p_i D \tag{4.33}$$

where K corresponds to the input power levels per phase, i and l correspond to the input power level for the ith instant at the lth phase, μ_i, n_i, and p_i correspond to the mean, efficiency, and the probability of the ith power level, respectively, and D depends on the total time considered. For an annual IEO, the total time considered $D = 8\,760$ hrs. Generally, while dealing with aging and degradation failures, the IEO is estimated for the first year of the PV system life cycle as it gives the IEO.

B. Expected energy output

The expected energy output (EEO) of a PV system is associated with the non-perfect reliability, and it is estimated by multiplying the ideal output of the generation with the availability of the system. Further, the total EEO is obtained by multiplying the sum of expected outputs at each power level with the probability of each power level.

The EEO for a central inverter system is estimated as

$$\text{EEO} = \sum_l \left[\sum_{i=1}^{K} \eta_i p_i D \sum_{j \in \{0,\, 1,\, 2,\, ...,\, n-1\}} f_j \mu_i p_{Aj} A_{I,ij} \left(f_j \mu_i,\, V_{DC,i} \right) \right] A_{DC} A_{AC} \quad (4.34)$$

where $f_j \mu_i$ is the projected inverter input power considering failures of the PV array, p_{Aj} corresponds to the probability of PV array's jth state, $A_{I,ij}$ corresponds to inverter availability for the ith input power level and the jth state of PV array, A_{DC} represents the availability of the DC disconnect, and A_{AC} corresponds to the availability of the AC subpanel. The value of f_j while estimating the inverter input power is obtained as a ratio that depends on the state of the PV array and number of homogenous strings (n) in a PV array.

Similarly, the EEO for a string inverter PV system is given as

$$\text{EEO} = \sum_l \left[\sum_{i=1}^{K} \eta_i p_i D \sum_{j = \{0,\, 1,\, 2,\, ...,\, n-1\}} f_j \mu_i p_{Aj} \left(\mu_{str,i},\, V_{DC,str,i} \right) \right] A_{AC} \quad (4.35)$$

where p_{Aj} is the state probability function that defines the power flow through the DC-side voltage and the string inverters. This function implicitly incorporates the failure risk of string inverters, which is a major difference between the EEO of the central and string inverters. Further, $\mu_{str,i}$ corresponds to the string inverter input power at the ith power level, and $V_{DC,str,i}$ represents the string inverter DC-side voltage.

C. Energy availability (A_e)

The energy availability is calculated as a ratio of the normalized EEO to IEO, which is given as

$$A_e = \frac{\text{EEO}}{\text{IEO}} \quad (4.36)$$

Generally, the IEO is a constant as it is estimated for the first year of the PV system life cycle.

4.3.3.2   Equivalent parameters for reliability of PV string

The time-based indices quantify the annual availability and unavailability time of the PV systems to justify their maintenance requirements.

A. Availability time (A_t)

The availability time A_t is the relative measure of the expected operating time for a PV system in a year under normal conditions. The availability time of central and string inverter PV systems with multiple phases are given in (4.37) and (4.38), respectively, as

$$A_{t,cen} = \prod_l \left[\sum_{i=1}^{K} p_i p_{A0} A_{I,i0} \left(f_0 \mu_i,\, V_{DC,i} \right) A_{DC} A_{AC} \right] \quad (4.37)$$

$$A_{t,str} = \prod_l \left[\sum_{i=1}^{K} p_i p_{A0} \left(\mu_{str,i},\, V_{Dc,\, str,\, i} \right) A_{AC} \right] \quad (4.38)$$

For a PV system that does not need any repair or replacement, the availability time is represented as a percentage of the time. It should be noted that the availability time for a PV system also includes the zero-power output during no solar insolation.

Further, the time unavailability is given as

$$U_t = 1 - A_t \tag{4.39}$$

Notably, the PV system unavailability includes the probability that it operates with parts of the PV string out of service or in different derated states. These derated states can be estimated using the SEM.

B. Available (H_{av}), derated (H_{dr}), and outage hours (H_{dw})

The operation of a PV system for fully available hours H_{av} is given as

$$H_{av} = A_t \cdot 8760 \tag{4.40}$$

Similarly, the average down time of the PV plant gives the outage hours H_{dw} of the system. H_{dw} for the central and string inverter PV systems are given by (4.41) and (4.42), respectively:

$$H_{dw,cen} = 8760 \cdot \prod_l \left[1 - \sum_{i=1}^{K} p_i \sum_{j \in \{0,\,1,\,2,\,...,\,n-1\}} p_{Aj} A_{1,ij} \left(f_j \mu_i, V_{DC,i} \right) A_{DC} A_{AC} \right] \tag{4.41}$$

$$H_{dw,str} = 8760 \cdot \prod_l \left[1 - \sum_{i=1}^{K} p_i \sum_{j \in \{0,\,1,\,2,\,...,\,n-1\}} p_{Aj} \left(\mu_{str,i}, V_{DC,str,\,i} \right) A_{AC} \right] \tag{4.42}$$

Further, the operation of the PV system in the derated state is given as

$$H_{dr} = 1 - H_{av} - H_{dw} \tag{4.43}$$

Hence, the time-based reliability indices are the major asset for the intelligent management of the PV system.

4.4 Results of site study for reliability analysis

The reliability analysis is carried out considering the 20-kW central and string PV inverter systems connected to the distribution network. The inverter efficiency curve is not considered for the test case as the DC voltage and inverter outputs are measured directly. Further, the required reliability parameters and the discrete probability model of annual power outputs are discussed in Tables 4.2 and 4.3 of Section 4.2.

4.4.1 Results of reliability indices

The reliability parameters in Tables 4.2 and 4.3 are used to obtain the reliability indices for the base case during the first year of service as shown in Table 4.4.

From the results, it is identified that both the systems have similar reliability indices. The central inverter system has a slight advantage in terms of fully available time, and the string inverter system is slightly advantageous in terms of energy

Table 4.4   Reliability indices for the base case during the first year of service

Energy-based indices	Central inverter system	String inverter system
EEO (MWh)	19.84	19.91
IEO (MWh)	19.95	19.95
A_e	0.98917	0.99173
Time-based indices	**Central inverter system**	**String inverter system**
A_t	0.91004	0.90827
H_{av} (hrs)	7956.062	7901.3334
H_{dr} (hrs)	803.9317	858.6634
H_{dw} (hrs)	0.0063	0.0032

availability. Due to multiple inverters in the string-based system, their failure frequency is significant, which affects the fully available time of the system. Further, the impact of outage in a string inverter only affects the string, whereas the central inverter outage affects all the strings. This justifies the relatively high energy availability for a string inverter.

4.4.2   Aging and degradation effects

The long-term performance analysis for a PV system must deal with aging and degradation effects. Hence, energy availability (A_e) and availability time (A_t) indices are calculated for both central and string systems for 25 years, as shown in Figure 4.15.

From the results in Figure 4.15, it is observed that the energy availability and availability time are sensitive to change in service age for both PV systems. In contrast to both indices, the relative sensitivity of A_e is smoother than A_t, as A_t remains insensitive for almost 15 years of the operating life and falls quickly to approach the mean life of the PV array, where A_e depicts a constant decrease. This indicates that the changes in the degradation of PV efficiency and aging can be identified easily by A_e, as A_t only indicates the influence of aging because of the indirect impact of PV degradation. This occurs due to the effect of inverter input power on the failure rate of components. Both A_e and A_t are very low at the end of life with A_t nearly equal to zero, indicating a high repair requirement. This indicates the dominance of aging failure as the PV system reaches the end of life.

4.4.3   PV risk assessment

4.4.3.1   Impact of temperature

The PV risk assessment for the impact on ambient temperature is generally related to the inverter systems and their installation site. The central inverter is installed in an electrical room with additional cooling facilities, and it is assumed that its temperature may vary between 0 °C and 40 °C as per the operating scenario and climatic conditions in New Delhi, India. Further, the string inverter systems are mounted outdoor near the PV strings, and it is assumed that its temperature may vary between 0 °C and 60 °C considering their direct exposure to environmental conditions. While

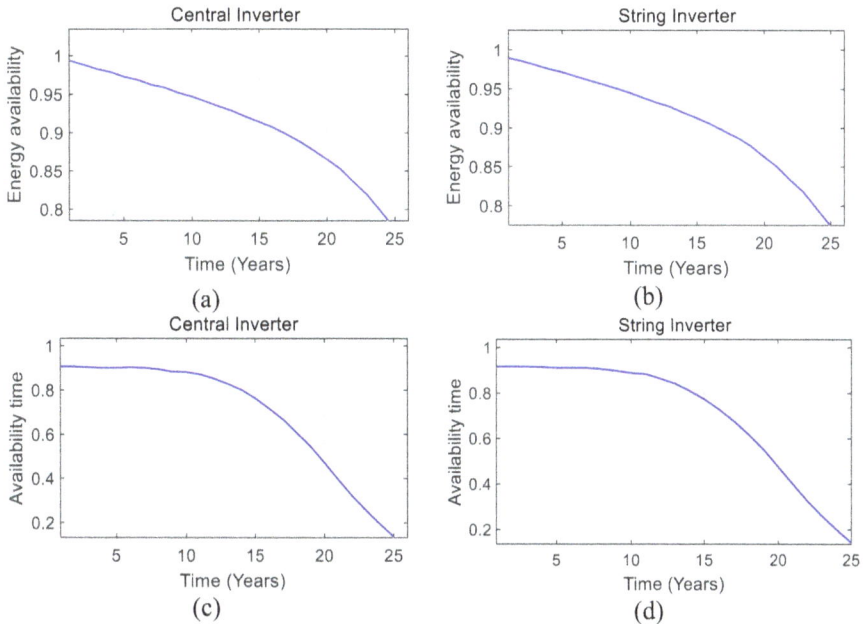

Figure 4.15 *Degradation effect on (a) energy availability of the central inverter, (b) energy availability of the string inverter, (c) availability time of the central inverter, and (d) availability time of the string inverter*

conducting the risk analysis, the temperatures below 0 °C are not considered due to low-performance risk. Besides, for a comparative analysis, the central inverter is subjected to temperatures between 40 °C and 60 °C and the sensitivity results are given in Table 4.5, Figure 4.16, and Tables 4.6 and 4.7.

From the results in Table 4.5, it is observed that the temperature rise has an equal impact on both systems. During the first service year, for an ambient temperature varying between 0 °C and 60 °C, the energy availability of the central inverter decreases from 99 percent to 96 percent. In contrast, the energy availability of string inverters varies from 99.2 percent to 99 percent only. This indicates that the string inverter is more tolerant of temperature changes seen from the energy availability perspective when compared with the central inverter under the same conditions. Similarly, at the end of life operation, i.e. for the 25th year, the energy availability results indicate the dominance of aging failures over the temperature impact. Further, the availability time during the first service year decreases with temperature for both systems with a drop from 90 percent to 89 percent for central inverter systems and 90 percent to 88 percent for string inverter systems. This indicates that additional maintenance is required while considering the temperature impact on PV systems.

From Figure 4.16, it is identified that the varying temperature influences the reliability of the PV system. For the experiment, it is assumed that the temperatures

Table 4.5   Impact of temperature on availability of PV inverter

Energy availability

Time (years)	Temperature 0		20		40		60	
	Central inverter	String inverter	Central inverter	String inverter	Central inverter	String inverter	Central inverter	String inverter
0	0.99	0.992	0.98	0.99	0.97	0.99	0.96	0.99
5	0.97	0.98	0.93	0.97	0.9	0.97	0.89	0.97
10	0.93	0.94	0.88	0.93	0.86	0.92	0.84	0.92
15	0.88	0.88	0.84	0.87	0.83	0.86	0.79	0.85
20	0.85	0.86	0.81	0.85	0.79	0.84	0.76	0.83
25	0.78	0.8	0.74	0.78	0.71	0.74	0.65	0.7

Availability time

Time (years)	Temperature 0		20		40		60	
	Central inverter	String inverter	Central inverter	String inverter	Central inverter	String inverter	Central inverter	String inverter
0	0.9	0.9	0.9	0.9	0.9	0.89	0.89	0.88
5	0.83	0.84	0.82	0.83	0.82	0.83	0.82	0.83
10	0.7	0.76	0.68	0.74	0.68	0.73	0.68	0.73
15	0.54	0.56	0.51	0.55	0.5	0.53	0.5	0.53
20	0.37	0.43	0.33	0.42	0.32	0.42	0.31	0.42
25	0.2	0.2	0.2	0.2	0.19	0.2	0.19	0.2

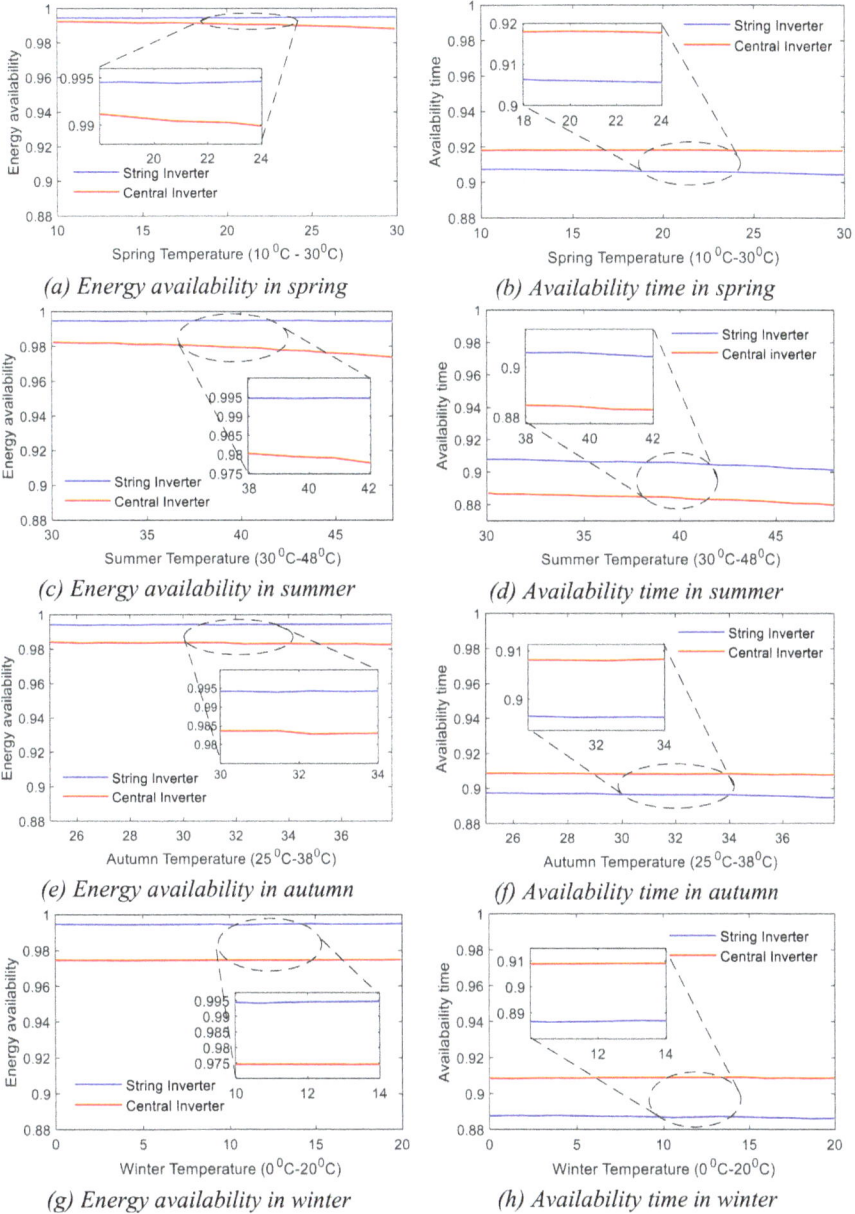

(a) Energy availability in spring

(b) Availability time in spring

(c) Energy availability in summer

(d) Availability time in summer

(e) Energy availability in autumn

(f) Availability time in autumn

(g) Energy availability in winter

(h) Availability time in winter

Figure 4.16   Impact of periodic temperature variations on the reliability of string and central inverters

vary between 10 °C and 30 °C in spring, 30 °C and 48 °C in summer, 25 °C and 38 °C in autumn, and between 0 °C and 20 °C in winter. From the results in Figure 4.16, it is observed that the string inverter system has a higher energy availability and lower

Table 4.6   Statistical parameters of temperature sensitivity test for energy availability index in different periods

Periodic division	Mean		Standard deviation	
	String	Central	String	Central
Spring	0.9931	0.9939	4.26×10^{-4}	6.97×10^{-3}
Summer	0.9928	0.9927	4.18×10^{-4}	7.62×10^{-3}
Autumn	0.9932	0.9953	4.02×10^{-4}	2.17×10^{-3}
Winter	0.9931	0.9967	4.07×10^{-4}	5.93×10^{-4}

availability time when compared with the central inverter system in spring, autumn, and winter seasons. This adheres to the reliability indices in Table 4.4, where the energy availability of string inverters is higher than the central inverters and the availability time of the central inverters is higher than the string inverters. However, during the summer period, both the energy availability and availability time of the string inverters are higher than the availability time of the central inverter.

Further, the energy availability and availability time in Figure 4.16 are analyzed using the mean and standard deviation of energy availability and availability time indices. From Table 4.6, it is identified that the string inverter dominates the central inverter over energy availability with high mean and low standard deviation. The higher mean value indicates the high energy production and the lower standard deviation indicates low sensitivity to change in temperature. This adheres to the condition that any failure in the string inverter blocks the power output of the only string, whereas the failure in central inverter blocks the complete power generation resulting in reduced energy availability. Similarly, the results in Table 4.7 indicate that the central inverter dominates the string inverter configuration over the availability time with high mean and low standard deviation. This adheres to the condition that the string inverter-based systems are prone to more failures due to the existence of more inverters and high redundancy. Further, the higher mean value and lower standard deviation for string inverter over availability time indicate the problems faced by the

Table 4.7   Statistical parameters of temperature sensitivity test for availability time index in different periods

Periodic division	Mean		Standard deviation	
	String	Central	String	Central
Spring	0.8976	0.9027	5.97×10^{-3}	3.92×10^{-3}
Summer	0.8992	0.8991	6.09×10^{-3}	6.17×10^{-3}
Autumn	0.9021	0.9042	5.68×10^{-3}	1.21×10^{-3}
Winter	0.9027	0.9061	5.57×10^{-3}	9.1×10^{-4}

central inverter due to high-power inputs during summer. This high-power phenomenon dominates the multiplicity of string inverters and results in more failures in the central inverter during summer.

4.4.3.2   Impact of solar insolation

The solar insolation defines the PV inverter input power, which is directly proportional to the power loss in IGBTs, diodes, and capacitors. This indicates that higher solar irradiance will result in high failure rates for the inverter. Further, to test the impact of solar insolation on PV systems, a sensitivity analysis is carried out by varying the PV inverter input power from 0.6 to 1.2 times of the standard input.

Figure 4.17 indicates that the energy availability of the central inverter system is highly vulnerable to variations in solar irradiance, especially during summer and spring. It decreases from 99.4 percent to 98 percent in summer and 99.3 percent to 98.3 percent in spring as the irradiance increases from 0.6 pu to 1.2 pu. Further, the energy and availability time of the string inverter system is less affected by the variation in irradiance for all the seasons as the system design evenly impacts the input power distribution. This indicates that the string inverters are less prone to the electrical stresses for an increase in solar irradiance level >1 pu. For a better understanding of the effect of solar irradiance on the energy and availability time of both inverters, the statistical parameters for results in Figure 4.17 are shown in Tables 4.8 and 4.9.

These statistical parameters correspond to the mean value and standard deviation of energy and availability time indices. From Table 4.8, it is observed that the string configuration is dominant in terms of energy availability index over the central configuration with a higher mean value and lower standard deviation. The higher mean values indicate high energy production whereas the lower standard deviation indicates lower sensitivity to temperature changes, except for the mean value in winter. This superiority is achieved by the string inverters as the failure of an inverter only affects the power generation of that specific string. Further, based on the statistical parameters for availability time in Table 4.9, it is observed that the string inverter is highly sensitive to solar irradiance, as the central inverter has a higher mean value.

4.4.3.3   Impact of capacitor equivalent series resistance

The capacitor ESR corresponds to the resistive part of the capacitor impedance that is largely observed in commonly used electrolytic capacitors for inverters. Generally, the increase in ESR results in higher hotspot temperatures making the capacitor more susceptible to failures. Further, to observe the sensitivity of inverter failures for changing ESR, a base case of the ESR variations by two times is generated and the results are given in Figure 4.18.

Figure 4.18 shows that the energy and availability time of string inverters are insensitive for ESR variations in all the seasons, whereas the central inverter system that is heavily equipped with cooling systems is insensitive to ESR variations especially in summer and spring. This is due to the even distribution of the input power

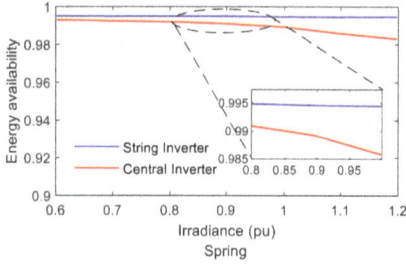

(a) Energy availability in spring

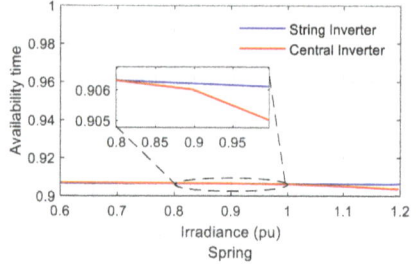

(b) Availability time in spring

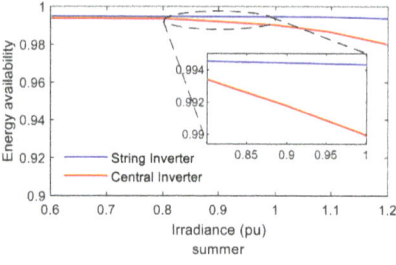

(c) Energy availability in summer

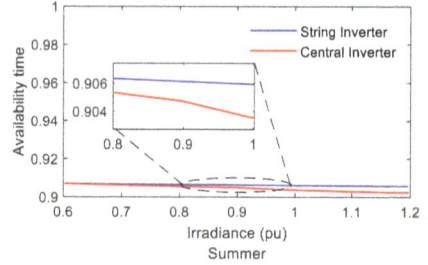

(d) Availability time in summer

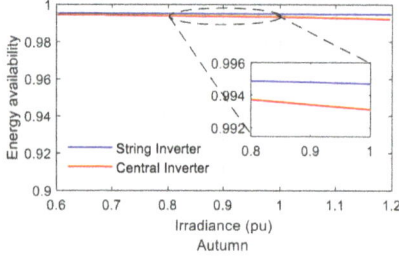

(e) Energy availability in autumn

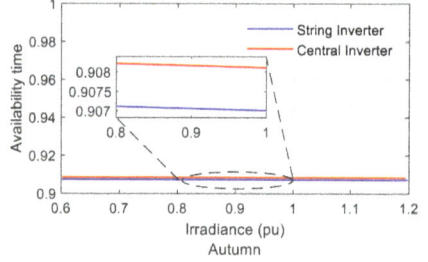

(f) Availability time in autumn

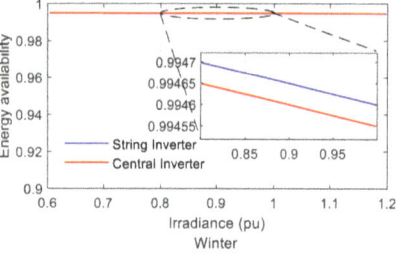

(g) Energy availability in winter

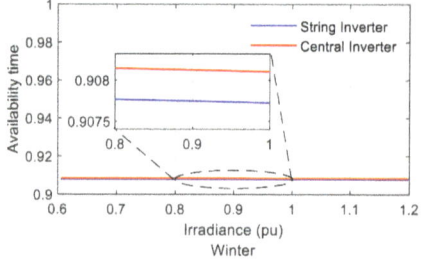

(h) Availability time in winter

Figure 4.17   Impact of solar irradiance variation on the reliability of string and central inverters

among the string inverters of each phase. Further, different studies have identified in [41, 42] that the RMS ripple current for string inverters is six times less than the RMS ripple current for the central inverter. This results in a significantly lower hotspot temperature, which further leads to lower capacitor failure rate and steady

Table 4.8 Statistical parameters of irradiance sensitivity test for energy availability index in different periods

Periodic division	Mean		Standard deviation	
	String	**Central**	**String**	**Central**
Spring	0.9932	0.9911	3.13×10^{-5}	3.47×10^{-3}
Summer	0.9927	0.9916	3.32×10^{-5}	4.68×10^{-3}
Autumn	0.9929	0.9921	1.83×10^{-5}	1.02×10^{-4}
Winter	0.9926	0.9937	5.71×10^{-6}	3.4×10^{-5}

energy and availability time for the string system. Moreover, the drawback of the central inverter system can be overcome by implementing an optimally rated capacitor at the design stage to ensure the system reliability.

4.4.4 Impact of the increased number of strings on PV system reliability

Further, to investigate the impact of risks due to the more distributed design of PV systems, the number of strings n in a PV array is varied by keeping the total output of the array at 7 kW. The results of the risk analysis are given in Table 4.10.

From the results in Table 4.10, it is observed that during the initial stages of service life, the energy availability and availability time of the central PV system are insensitive to an increase in the number of strings n. This is because the string failure rate reduces with several panels, but more contingencies occur with increasing n. These effects create an offset at the beginning of the PV system life cycle. As the PV system proceeds to the end of life, the energy availability and availability time decrease very quickly for an increased number of strings due to the dominance effect of the aging of components. In the case of a string inverter-based PV system, it is observed that the energy availability slightly decreases during the first 10 years of the service life with an increase in n. Similarly, the availability time drops to a

Table 4.9 Statistical parameters of irradiance sensitivity test for availability time index in different periods

Periodic division	Mean		Standard deviation	
	String	**Central**	**String**	**Central**
Spring	0.906	0.9062	1.3×10^{-4}	1.41×10^{-3}
Summer	0.9058	0.906	2.1×10^{-4}	2.97×10^{-3}
Autumn	0.9075	0.9082	4.99×10^{-5}	2.71×10^{-4}
Winter	0.9078	0.9081	1.17×10^{-5}	1.47×10^{-5}

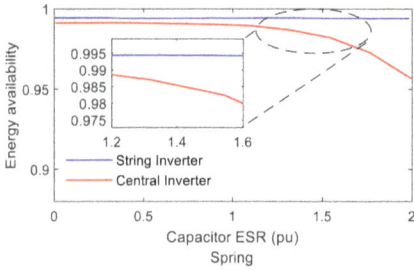

(a) Energy availability in spring

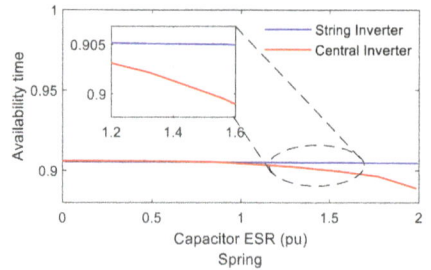

(b) Availability time in spring

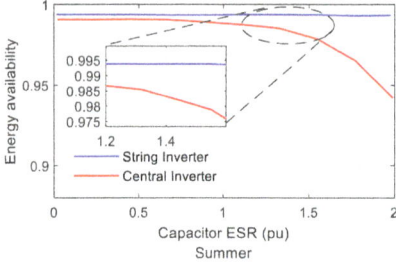

(c) Energy availability in summer

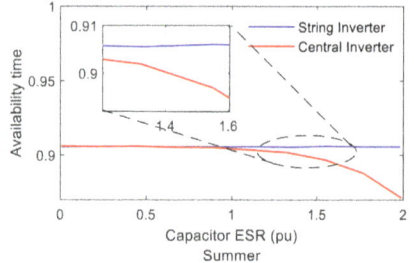

(d) Availability time in summer

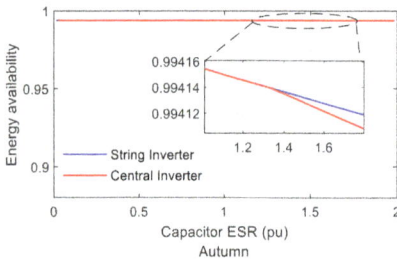

(e) Energy availability in autumn

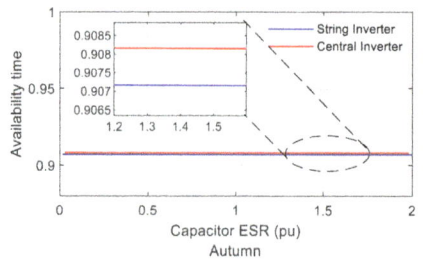

(f) Availability time in autumn

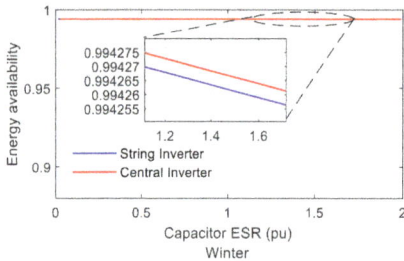

(g) Energy availability in winter

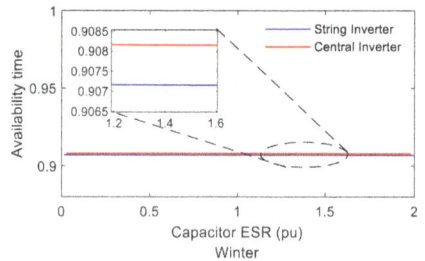

(h) Availability time in winter

Figure 4.18   Impact of capacitor ESR on the reliability of string and central inverters

Table 4.10   Impact of the increased number of strings on the availability of PV inverters

Energy availability

Time (years)	Number of strings 4		8		12		16	
	Central inverter	String inverter	Central inverter	String inverter	Central inverter	String inverter	Central inverter	String inverter
0	0.98	0.98	0.98	0.98	0.97	0.98	0.96	0.98
5	0.91	0.94	0.89	0.94	0.88	0.94	0.88	0.93
10	0.88	0.91	0.87	0.91	0.87	0.89	0.83	0.88
15	0.79	0.88	0.76	0.87	0.73	0.87	0.71	0.85
20	0.76	0.83	0.73	0.78	0.71	0.75	0.68	0.7
25	0.73	0.81	0.71	0.76	0.68	0.71	0.6	0.62

Availability time

Time (years)	Number of strings 4		8		12		16	
	Central inverter	String inverter	Central inverter	String inverter	Central inverter	String inverter	Central inverter	String inverter
0	0.88	0.88	0.88	0.86	0.88	0.86	0.88	0.86
5	0.85	0.86	0.84	0.85	0.84	0.85	0.84	0.85
10	0.73	0.81	0.71	0.78	0.67	0.75	0.6	0.73
15	0.68	0.69	0.62	0.63	0.56	0.58	0.48	0.54
20	0.52	0.51	0.43	0.46	0.35	0.4	0.26	0.32
25	0.24	0.2	0.18	0.12	0.11	0.07	0	0

Table 4.11   Impact of panel failure rate on availability of PV inverters

Energy availability

Time (years)	Panel failure rate (pu) 0		1		2	
	Central inverter	String inverter	Central inverter	String inverter	Central inverter	String inverter
0	0.97	0.98	0.97	0.97	0.96	0.97
5	0.94	0.93	0.92	0.91	0.91	0.9
10	0.89	0.87	0.88	0.85	0.85	0.83
15	0.82	0.83	0.8	0.8	0.79	0.78
20	0.8	0.78	0.78	0.77	0.76	0.75
25	0.78	0.76	0.77	0.75	0.75	0.75

Availability time

Time (years)	Number of strings 0		1		2	
	Central inverter	String inverter	Central inverter	String inverter	Central inverter	String inverter
0	0.87	0.88	0.83	0.83	0.8	0.8
5	0.86	0.85	0.83	0.81	0.8	0.8
10	0.79	0.79	0.76	0.75	0.7	0.73
15	0.67	0.62	0.56	0.53	0.51	0.44
20	0.61	0.53	0.49	0.41	0.31	0.33
25	0.34	0.3	0.16	0.17	0	0

significant value with increasing average repair time for an increase in n. This indicates that the maintenance requirements for increasing n in string inverters elevate quickly, resulting in higher maintenance costs.

4.4.5   Impact of panel failure rate on PV system reliability

The effect of PV panel failure rate λ_p on the reliability of the PV system is assessed through sensitivity analysis as shown in Table 4.11.

From the results in Table 4.11, it is observed that the failure rate has a similar effect on energy availability and availability time of both central and string connected PV systems. This is due to the high number of series-connected modules in each string. Further, it can be specified that the availability time is more sensitive to the failure rate than the energy availability. Notably, the repair time for the PV panel r_p has similar sensitivity characteristics with the failure rate of the PV panel. This is because the PV panel availability is equal to $\frac{\lambda_p r_p}{(1+\lambda_p r_p)}$, where both the variables λ_p and r_p are exchangeable.

4.5   Summary

This chapter discussed a case study to evaluate the reliability of a PV system. The developed approach quantified the effects of different operational conditions on the reliability of PV arrays, inverters, and overall PV system. Further, two different PV system configurations developed on 20-kW grid-connected central and string inverter systems are identified to perform the reliability analysis by adopting SEMs. Besides, the major contributions include risk analysis for periodic impacts on the PV systems and effect analysis of aging failure models and different operational conditions. These aspects of the developed approach and their implementation with actual PV systems provide efficient design options for PV systems and valuable information for operation and maintenance.

References

[1] National Institution for transforming India. *Report of the expert group on 175 GW*. New Delhi; 2015.

[2] Renewable Energy Project Monitoring Division. Plant wise details of all India renewable energy projects. 6/3/2019/1410. New Delhi, India: Central electricity authority, Ministry of Power, Government of India; 2020. Available from https://cea.nic.in/wp-content/uploads/2020/04/Plant-wise-details-of-RE-Installed-Capacity-merged.pdf [Accessed 15th May 2021].

[3] Alagh Y.K., IEA. 'India 2020 policy energy review'. *Journal of Quantitative Economics: Journal of the Indian Econometric Society*. 2006;**4**(1):1–14.

[4] Cronin J., Anandarajah G., Dessens O. 'Climate change impacts on the energy system: A review of trends and gaps'. *Climatic Change*. 2018;**151**(2):79–93.

[5] Drück H., Pillai R.G., Tharian M.G., Majeed A.Z. (eds.). *Green Buildings and Sustainable Engineering*. Singapore: Springer Singapore; 2019.

[6] Indian Met Department. *Ever Recorded Maximum Temperature, Minimum Temperature and 24 Hours Heaviest Rainfall upto 2010*. Pune; 2010.

[7] Häberlin H. 'Normalized representation of energy and power of PV systems'. *Photovoltaics: System Design and Practice*. 1st edn. Chichester, UK: John Wiley & Sons, Ltd; 2012. pp. 487–506. Available from https://onlinelibrary.wiley.com/doi/book/10.1002/9781119976998.

[8] Sungrow. SG15KTL-M/SG20KTL-M Multi-MPPT string inverter for 1000 Vdc system. Version 13. 2019. Available from https://www.europe-solarstore.com/download/Sungrow/SUNGROW_SG15_20KTL-M_V13_Datasheet.

[9] Sungrow. 'SG1250UD/SG1500UD outdoor inverter for 1000 Vdc system'. 2019.

[10] Saha K. 'Planning and installing photovoltaic system: a guide for installers, architects and engineers'. *International Journal of Environmental Studies*. 2014:1–2.

[11] Formica T.J., Khan H.A., Pecht M.G. 'The effect of inverter failures on the return on investment of solar photovoltaic systems'. *IEEE Access*. 2017;**5**:21336–43.

[12] Flicker J., Kaplar R., Marinella M., Granata J. 'PV inverter performance and reliability: what is the role of the bus capacitor?' *2012 IEEE 38th Photovoltaic Specialists Conference (PVSC) PART*. 2012;**2**:1–3.

[13] Harms J.W. Revision of MIL-HDBK-217, Reliability prediction of electrical equipment. Annual Symposium on Reliability and Maintainability (RAMS); San Jose, CA, USA, 25-28 January 2010; 1991. pp. 1–3.

[14] FIDES Group. Reliability methodology for electronics systems. FIDES guide 2009 Edition A. France: Union Technique de l'Electricité; 2010. pp. 1–465. Available from https://www.fides-reliability.org/en [Accessed 15th May 2021].

[15] 'Jonathan S. (Ed.). *Reliability Characterisation of Electrical and Electronic Systems*. Elsevier'. 2015.

[16] Kurukuru V.S.B., Haque A., Khan M.A., Tripathy A.K. 'Reliability analysis of silicon carbide power modules in voltage source converters'. 2019 International Conference on Power Electronics, Control and Automation (ICPECA); 2019. pp. 1–6.

[17] Department of Defense of the USA. 'Reliability prediction of electronic equipment'. *Military Handbook (MIL-HDBK)*. 1991;**217F**:205.

[18] Zhou D., Blaabjerg F., Lau M., Tonnes M. Thermal profile analysis of doubly-fed induction generator based wind power converter with air and liquid cooling methods. 2013 15th European Conference on Power Electronics and Applications (EPE); 2013. pp. 1–10.

[19] Shahzad M., Bharath K.V.S., Khan M.A., Haque A. 'Review on reliability of power electronic components in photovoltaic inverters'. 2019 International Conference on Power Electronics; 2019. pp. 1–6.

[20] Liu X., Li L., Das D., Pecht M. 'An electro-thermal parametric degradation model of insulated gate bipolar transistor modules'. *Microelectronics Reliability*. 2020;**104**(46(9)):113559.

[21] Infineon Technologies AG. IGBT selection guide-Common IGBT applications and topologies. B133-I0528-V1-7600-EU-EC. 9500 Villach, Austria: Infineon Technologies Austria AG; 2018. pp. 1–8. Available from https://www.infineon.com/cms/austria/en/ [Accessed 15th May 2021].

[22] Clemente S. A simple tool for the selection of IGBTs for motor drives and UPSs. Proceedings of *1995* IEEE Applied Power Electronics Conference and Exposition (APEC); 1995. pp. 755–64.

[23] Flicker J.D. Capacitor reliability in photovoltaic inverters. Unlimited Release. Albuquerque, New Mexico 87185 and Livermore, California 94550: Sandia National Laboratories; 2012. pp. 1–78. Available from http://www.ntis.gov/help/ordermethods.asp?loc=7-4-0#online [Accessed 15th May 2021].

[24] Nelson J.J., Venkataramanan G., El-Refaie A.M. 'Fast thermal profiling of power semiconductor devices using Fourier techniques'. *IEEE Transactions on Industrial Electronics*. 2006;**53**(2):521–9.

[25] Ton D B.W. 'Summary report on the DOE high-tech inverter workshop'. *US Department of Energy*. 2005;**2005**.

[26] U.S. Department of Energy. Workshop Summary Report: R & D for Dispatchable Distributed Energy Resources at Manufacturing Sites; 2016. p. 32.

[27] Harb S., Balog R.S. Reliability of candidate photovoltaic module-integrated-inverter topologies. *2012 Twenty-Seventh Annual IEEE Applied Power Electronics Conference and Exposition (APEC)*; 2012. pp. 898–903.

[28] Li S. *Condition-dependent risk assessment of large-scale grid-tied photovoltaic power systems*. (University of Connecticut) [Master's Thesis] Storrs, CT 06269, United States, University of Connecticut Graduate School. 2012. Available from https://opencommons.uconn.edu/gs_theses/370 [Accessed 15th May 2021].

[29] Warren C.A., Saint R. 'IEEE reliability indices standards'. *IEEE Industry ApplicationsMagazine*. 2005;**11**(1):16–22.

[30] Moharil R.M., Kulkarni P.S. 'Reliability analysis of solar photovoltaic system using hourly mean solar radiation data'. *Solar Energy*. 2010;**84**(4):691–702.

[31] Cristaldi L., Khalil M., Faifer M. 'Markov process reliability model for photovoltaic module failures'. *Acta Imeko*. 2017;**6**(4):121.

[32] Sayed A., El-Shimy M., El-Metwally M., Elshahed M. 'Reliability, availability and maintainability analysis for grid-connected solar photovoltaic systems'. *Energies*. 2019;**12**(7):1213.

[33] Hu R., Mi J., Hu T., Fu M., Yang P. 'Reliability research for PV system using BDD-based fault tree analysis'. 2013 International Conference on Quality; 2013. pp. 359–63.

[34] Zhang P., Li W., Li S., Wang Y., Xiao W. 'Reliability assessment of photovoltaic power systems: review of current status and future perspectives'. *Applied Energy*. 2013;**104**(1):822–33.

[35] Ghahderijani M.M., Barakati S.M., Tavakoli S. 'Reliability evaluation of stand-alone hybrid microgrid using sequential Monte Carlo simulation'. 2012 Second Iranian Conference on Renewable Energy and Distributed Generation; 2012. pp. 33–8.

[36] Billinton R., Hua Chen., Ghajar R. 'A sequential simulation technique for adequacy evaluation of generating systems including wind energy'. *IEEE Transactions on Energy Conversion*. 1996;**11**(4):728–34.

[37] Haque A., Bharath K.V.S., Khan M.A., Khan I., Jaffery Z.A. 'Fault diagnosis of photovoltaic modules'. *Energy Science & Engineering*. 2019;**7**(3):622–44.

[38] Kurukuru V.S.B., Blaabjerg F., Khan M.A., Haque A. 'A novel fault classification approach for photovoltaic systems'. *Energies*. 2020;**13**(2):308.

[39] Ahmad S., Hasan N., Bharath Kurukuru V.S., Ali Khan M., Haque A. 'Fault classification for single phase photovoltaic systems using machine learning techniques'. 2018 8th IEEE India International Conference on Power Electronics (IICPE); 2018. pp. 1–6.

[40] Kurukuru V.S.B., Haque A., Khan M.A., Tripathy A.K. 'Fault classification for photovoltaic modules using thermography and machine learning techniques'. 2019 International Conference on Computer and Information Sciences (ICCIS); 2019. pp. 1–6.

[41] Bianchi N., Dai Pre M. 'Active power filter control using neural network technologies'. *IEE Proceedings-Electric Power Applications*. 2003;**150**(2):139–45.

[42] Martins D.C. 'Analysis of a three-phase grid-connected PV power system using a modified dual-stage inverter'. *ISRN Renewable Energy*. 2013;**2013**(5):1–18.

Chapter 5

Control strategy for grid-connected solar inverters

Zhongting Tang[1] and Yongheng Yang[2]

Abstract

As an essential interface between the photovoltaic (PV) panels and the utility grid, solar PV inverters are responsible for converting intermittent solar energy to meet the utility grid requirement, where the inverter output should be synchronized with the grid voltage in terms of phase frequency and amplitude. In addition, considering system cost, conversion efficiency, power quality, and reliability of the grid-connected PV system, the control strategy of solar inverters should be carefully designed. Regarding grid-connected solar inverters, the basic control strategies include a maximum power point tracking (MPPT) algorithm (i.e., increasing efficiency and maximizing the energy harvesting), a DC-link voltage control, and a grid-connected current control (i.e., responsible for the power injection and current quality). In this chapter, the model of PV modules and a few typical MPPT methods are briefly introduced. Then, the DC-link voltage control and grid-connected current control are presented for the single-phase and three-phase solar inverters, respectively.

5.1 Introduction

5.1.1 Demands for grid-connected solar inverters

The solar systems generally consist of PV panels (i.e., module, strings, or arrays), power electronics (i.e., solar inverters, used for solar energy harvesting in grid-connected applications), and electric grid, as shown in Figure 5.1. The continuous development of solar power-generation systems can provide more clean energy, yet it also poses threats in terms of stability and economic energy management to the utility grid [1] (highly fluctuating due to intermittency). Therefore, to achieve a grid-friendly solar PV system with good performances of reliability, efficiency,

[1]Aalborg University, Aalborg, Denmark
[2]Zhejiang University, Hangzhou, China

and stability, the requirements for grid-connected solar inverters become more strict and flexible in many codes, exemplified as in [2]. As shown in Figure 5.1, the demands for a grid-connected solar system have three aspects. Referring to the PV side, maximum power harvesting and good maintenance of the PV panels should be ensured to enhance high energy utilization as well as a long lifetime. For the grid side, the requirements of grid-supporting capabilities have also been increased in addition to good power quality, voltage or frequency regulation as well as the abnormal grid voltage protection and recovery. As the key interface, the grid-connected solar inverter not only should consider most of the above requirements in both PV and grid sides but also has great demands for high-efficiency conversion and proper temperature management [3]. The main goal is to increase the reliability of solar inverters, producing the most energy with the least cost. In addition, communication is specifically required nowadays for intelligent and cooperative smart solar systems.

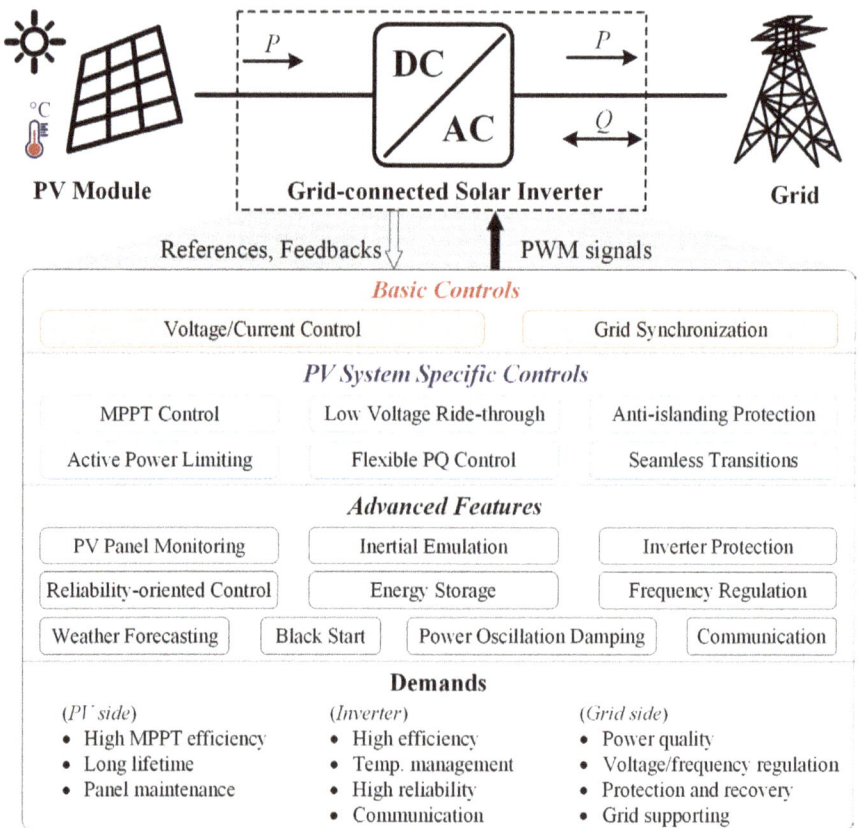

Figure 5.1   General control structure for the grid-connected solar PV inverter

5.1.2 General controls

To meet the above demands, reliable and efficient controls should be performed on solar PV inverters. With the drastically increasing solar capacity, more advanced features have been required in addition to basic controls of grid-connected inverters and PV-system-specific controls, as detailed below.

1. ***Basic controls***—As presented in Figure 5.1, the common basic controls of grid-tied inverters include voltage or current controls and grid synchronization schemes. In addition to improving the efficiency and reliability of the grid-connected solar PV inverters, another control objective is to have good steady-state and dynamic performances, achieve good power quality (e.g., a low total harmonic distortion level as grid requirements [4, 5]), and synchronize well with the grid voltage.

2. ***PV-system-specific controls***—As known, power generation from solar PV systems is highly dependent on the weather or climate conditions [1]. Thus, inverters applied in solar PV systems should have many specific controls, such as the MPPT control (i.e., MPPT at any solar radiation and temperature conditions) [6], active power limiting (i.e., alleviating the effect of the power fluctuation), anti-islanding protection, and fast recovery (i.e., grid resilience). In addition, the specific controls for the wind systems, e.g., low-voltage ride-through, are now mandatory in PV systems, as presented in IEEE 1547-2018 [2]. Correspondingly, more flexible PQ (i.e., active power P and reactive power Q) control is required to provide grid support [7], e.g., frequency control through active power and voltage control by reactive power.

3. ***Advanced features***—Since reliability and grid-supporting capability are emphasized more and more in today's grid-connected solar systems (i.e., high PV-capacity integration), many advanced features should be considered in the solar PV inverter controls. For instance, system condition monitoring and maintenance of PV panels achieve a long lifetime and high reliability. For a good grid-supporting performance of grid-connected solar PV inverters, delta power production control, artificial and virtual inertia controls, black start for enhancing the grid resilience, power oscillation damping, and implementing virtual synchronous generators by energy storage have been adopted in [8–10]. In addition, reliability-oriented controls (e.g., power-limiting control with weather forecasting and junction temperature control [11]) can also be employed to enhance the reliability and lifetime of the solar PV system (both the PV panels and solar inverters).

Nevertheless, those PV special controls and advanced features can be achieved by modifying the universal controls of solar inverters, e.g., MPPT control and current or voltage controls, which is the focus of this chapter. Generally, an MPPT algorithm is integrated into either the DC voltage or the AC output current control in single-stage solar inverters and implemented by the DC–DC converter in two-stage solar inverters [3]. Additionally, the DC-link voltage control and grid-connected

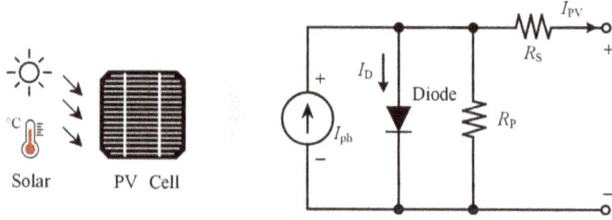

Figure 5.2   Equivalent circuit of a PV cell model

current control are well implemented on the DC–AC conversion stage [12]. In this chapter, first, the PV model, as well as some conventional MPPT algorithms, is depicted. Then, the modeling and controller design of the DC-link voltage and grid-connected current control will be introduced for both single-phase and three-phase inverters. In the end, the controller design will be demonstrated and verified with two case studies. One is a two-stage three-phase inverter in the *dq*-reference frame with proportional-integral (PI) controllers, including the current controller, DC-link voltage controller, and active power and reactive power controller. Another is the proportional-resonant (PR) controller for a single-phase inverter.

5.2   MPPT control

5.2.1   Modeling of PV panels

It is known that the PV panel model and its characteristics under different tempera-tures and irradiance levels are of significance to the PV system analysis (i.e., espe-cially the MPPT control design) [13]. As shown in Figure 5.2, a single-diode model of the PV cell is presented, where the output characteristics can be expressed as

$$I_{PV} = I_{ph} - I_O \left[\exp\left(\frac{V_{PV} + R_S I_{PV}}{AKT/q} \right) - 1 \right] - \frac{V_{PV} + R_S I_{PV}}{R_P} \tag{5.1}$$

where I_{PV} and V_{PV} are the PV cell output current and voltage, A is the ideality factor of the p-n junction, K is the Boltzmann's constant ($1.3806503 \times 10^{-23}$ J/K), T is the cell temperature in Kelvin (for simplicity, in practice, the ambient temperature is considered), q is the charge of an electron (1.6×10^{-19} C), R_S and R_P are the PV cell series and shunt resistances (i.e., $R_S \ll R_P$), respectively, and I_{ph} is the photocurrent, depending on both the solar irradiance G and the ambient temperature T as

$$I_{ph}(G, T) = \left(I_{scn} + K_i (T - T_n) \right) \frac{G}{G_n} \tag{5.2}$$

where I_{scn} and G_n are the nominal short-circuit current and the nominal solar irradi-ance (i.e., $G_n = 1\,000$ W/m²) under the nominal cell temperature T_n (i.e., $T_n = 25$ °C $= 298.15$ K), respectively, and K_i is the current temperature coefficient. In (5.1), the diode saturation current I_O is related to the ambient temperature according to

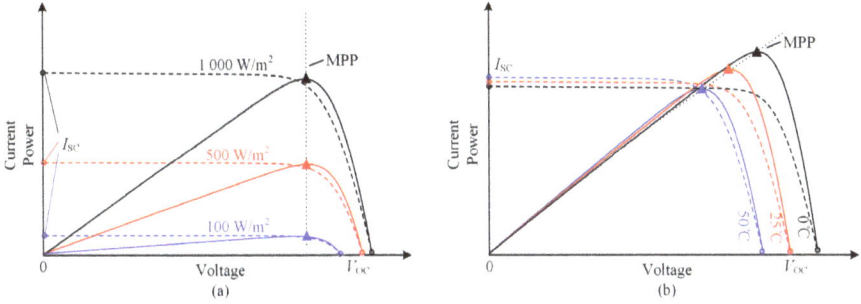

Figure 5.3 I–V *and* P–V *characteristics of a PV cell: (a) different solar irradiance levels at 25 °C and (b) different temperatures at 1 000 W/ m²*

$$I_O\left(T\right) = \frac{I_{scn} + K_i\left(T - T_n\right)}{\exp\left(\dfrac{V_{ocn} + K_v\left(T - T_n\right)}{aV_t\left(T\right)}\right) - 1} \tag{5.3}$$

in which V_{ocn} is the nominal open-circuit voltage under T_n, K_v is the voltage tempera-ture coefficient, a is the diode ideality factor, and V_t is the thermal voltage.

According to (5.1)–(5.3), the irradiance and ambient temperature highly affect I_{ph} and I_O and, then, the PV output power. The current–voltage (I–V) and power–voltage (P–V) characteristics are shown in Figure 5.3, which indicate that the maximum power point (MPP) is varying with ambient conditions. Thus, the solar inverter should integrate proper controllers to track the MPP under different weather conditions.

5.2.2 MPPT algorithm

As demonstrated in Figure 5.3, the P–V characteristics curves are always of "hill" nature, where the hilltop represents the MPP. To track the MPP under different ambient conditions, many MPPT algorithms have been proposed, such as the Constant Voltage (CV), the Perturb and Observe (P&O), the Incremental Conductance (INC), the Sliding Control method, the fuzzy logic, and the neural network [14]. Among them, the P&O and INC, which are also known as "hill-climbing" methods, are widely used due to their good performance in terms of simple implementation, fewer sensor requirements, and relatively high efficiency.

In this chapter, the P&O MPPT algorithm is exemplified to clarify the principle. Figure 5.4 depicts the flowchart of the P&O MPPT algorithm, where $V(k)$ and $I(k)$ are the PV output voltage and current at the time instant of k, and the output power is $P(k) = V(k) \times I(k)$, correspondingly. As shown in Figure 5.4, the P&O operation can be sum-marized as:

1. When $dP/dV > 0$ (i.e., $dP = \Delta P$ and $dV = \Delta V$), the P&O MPPT algorithm acts in the forward perturbation, i.e., increasing the reference voltage of the PV panel ($V_{ref} = V(k) + V_{step}$);

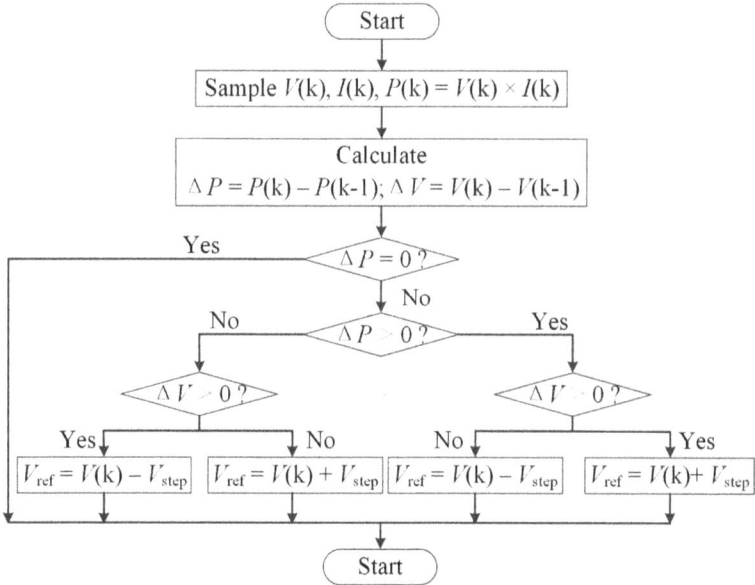

Figure 5.4   Flowchart of the P&O MPPT algorithm

2. When $dP/dV < 0$, the perturbation direction of the algorithm is reversed, where the reference voltage of the PV panel decreases (i.e., $V_{ref} = V(k) - V_{step}$, in which V_{step} is the perturbation step-size).

Obviously, the larger the V_{step} is, the more rapid the tracking is, yet the larger the power oscillation is. Therefore, a modified P&O MPPT algorithm with a variable step-size can be adopted. According to different ΔP, the modified MPPT algorithm adopts different perturbation step-sizes V_{step} to achieve a trade-off between the tracking speed and the power oscillation [15].

As mentioned in Section 5.1.2, the MPPT algorithm should be integrated into the control of the single-stage inverter or implemented only by the DC–DC converter of a two-stage system [12]. In this chapter, simulation results of a 4-kW two-stage grid-connected solar inverter with a trapezoidal solar irradiance profile are shown in Figure 5.5, which compares the performance of the traditional P&O MPPT algorithm in Figure 5.4 and the variable step-size P&O MPPT algorithm in [15]. As presented in Figure 5.5, the traditional P&O MPPT algorithm can have a rapid tracking of the MPP, and the power oscillation appears near the MPP. By comparison, the MPPT algorithm with variable perturbation step-size can achieve rapid tracking as well as small power variations. It should be mentioned that the above algorithms (e.g., CV, P&O, and INC) are not suitable for tracking the global MPP of multiple PV panels under partial shading conditions [6, 14]. The prior-art global MPPT algorithms have been compared in [16], such as the particle swarm optimization algorithm [17] and the Cuckoo search algorithm [18].

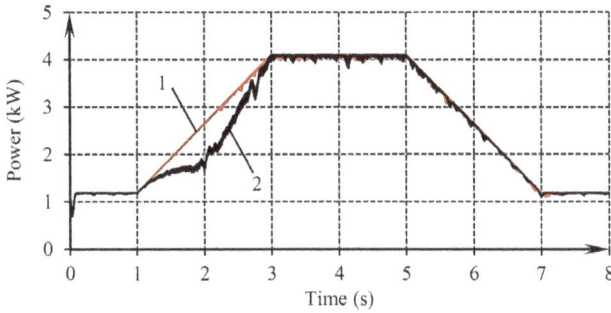

Figure 5.5   Performance comparison of the modified (line 1) and conventional (line 2) P&O MPPT algorithm (ambient temperature 25 °C)

5.3   Solar inverter control

To simplify the control analysis of the PV system, a two-stage inverter system is shown in Figure 5.6, which includes the PV array, the DC–DC optimizer, the grid-connected inverter, the passive components, and the grid. Referring to control tasks, the first stage (i.e., the DC–DC converter) aims to implement the MPPT algorithm, while the second stage focuses on transferring the extracted power from the DC–DC converter to the grid with low energy losses [19–21]. In addition, the inverter should have the performance in terms of good dynamics, robust synchronization, high power quality, and grid-supporting capability (i.e., reactive power injection, islanding protection, low-voltage ride-though capability, etc.) [22–24]. Therefore, the DC-link voltage and the grid current should be regulated effectively to achieve those goals.

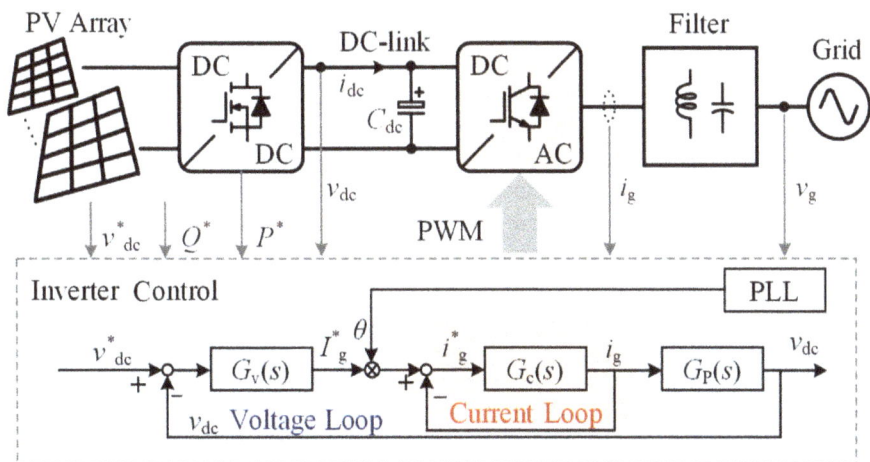

Figure 5.6   Dual-loop control for a two-stage grid-connected solar inverter

As demonstrated in Figure 5.6, the typical double closed-loop control for the two-stage grid-connected solar inverter contains a current inner loop control and a voltage outer loop control [12, 25]. As shown in Figure 5.6, v_{dc}, i_{dc}, v_g, and i_g represent the sampling value of the DC-link voltage, the DC current, the grid voltage, and the grid current, respectively. v^*_{dc} and i^*_g are the reference DC-link voltage and the reference grid current. I^*_g is the amplitude of i^*_g. $G_v(s)$, $G_c(s)$, and $G_p(s)$ are the DC-link voltage controller, the current controller, and the plant. θ is the phase angle, which is achieved by employing a PLL on the grid voltage v_g. The DC-link voltage controller $G_v(s)$ generates the grid current amplitude I^*_g by regulating the error of the reference DC-link voltage v^*_{dc} and the sampling one v_{dc}. By multiplying I^*_g and the phase-locked loop (PLL)-synchronized signal θ of the grid voltage v_g, the grid current reference i^*_g can be obtained. Then, the current control $G_c(s)$ generates the pulse width modulation (PWM) signals for the plant $G_p(s)$ to achieve the zero-tracking of the grid current i_g. The following will present the modeling and design of the grid current inner loop control and the DC-link voltage outer loop controller (i.e., including the PQ control) for both the single-phase and three-phase inverters, as shown in Figure 5.7, where an L-type filter (i.e., the inductance and its equivalent resistance are depicted) is adopted in both solar inverters.

5.3.1   Reference frame transformation

According to IEEE 1547 Std. [2], the solar inverters should have flexible power control capability (i.e., active power control and reactive power control) to achieve grid supporting. The PQ control based on the PQ theory is a way to achieve this goal, where the PQ theory needs the reference frame transformations in both the single-phase and three-phase inverters [26, 27]. Therefore, the following will depict the reference frame transformations, which include the Clarke and Park transformations (i.e., the AC control variables, including voltages and currents, will be transformed into two DC quantities). The reference frame transformations and the PQ controller can be applied to both single- and three-phase solar inverter system [28].

A. Clarke transformation (abc→ αβ)

The circuit diagram of a three-phase inverter is presented in Figure 5.7b, where v_{ga}, v_{gb}, v_{gc} and i_{ga}, i_{gb}, and i_{gc} represent the three-phase grid voltages and grid currents. Assuming that the grid-connected solar system is balanced when the inverter employs a three-phase inverter, the three-phase system can be transferred to a two-phase system on the stationary reference frame (i.e., the $\alpha\beta$-reference frame). The transformation can be expressed as

$$\begin{bmatrix} x_\alpha \\ x_\beta \end{bmatrix} = \frac{2}{3}\begin{bmatrix} 1 & \dfrac{-1}{2} & \dfrac{-1}{2} \\ 0 & \dfrac{\sqrt{3}}{2} & -\dfrac{\sqrt{3}}{2} \end{bmatrix}\begin{bmatrix} x_a \\ x_b \\ x_c \end{bmatrix} \tag{5.4}$$

in which x_a, x_b, and x_c are the control variables (e.g., voltages or currents of the system) and x_α and x_β are transferred variables in the $\alpha\beta$-reference frame [29].

(a)

(b)

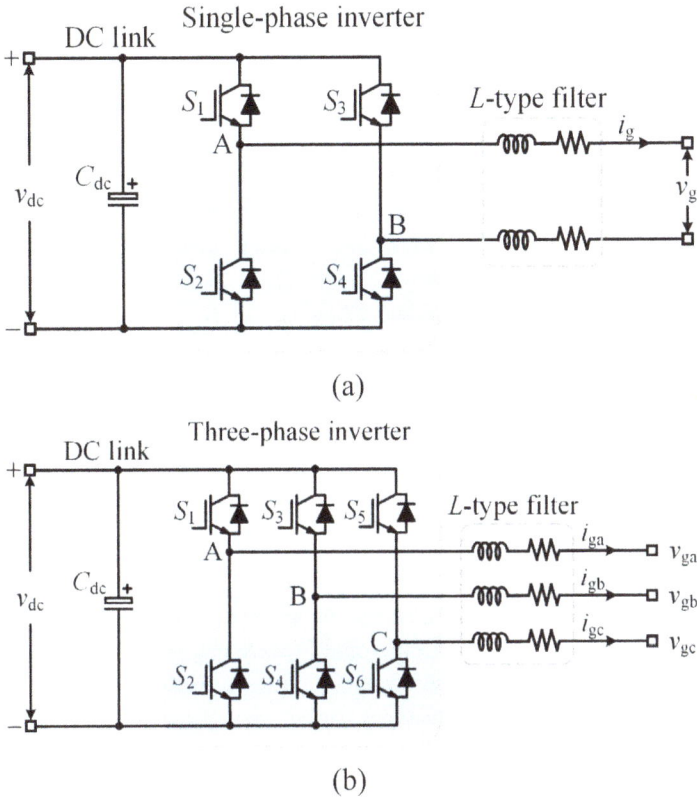

Figure 5.7   *Circuit diagram of two general solar inverters with an L-type filter:*
(a) single-phase inverter and (b) three-phase inverter

B. Park transformation (αβ→ dq)

The main advantage of the flexible control based on the PQ theory in both the single-phase and three-phase inverter is to directly synthesize the power references. To simplify the control algorithm, the traditional PI controller is enough to have good steady-state and dynamic performance. However, the variables in the $\alpha\beta$-reference frame in (5.4) are still AC variables rotating at the grid frequency, where the grid angle frequency is ω. In that case, the PI controller for the $\alpha\beta$ components (i.e., being period variable) has poor performance in terms of zero-error tracking [30]. To tackle this issue, the $\alpha\beta$ components can be transferred to the dq-reference frame (i.e., the synchronous reference frame), which is known as the Park transformation [31]. The $\alpha\beta \rightarrow dq$ transformation is given as

$$\begin{bmatrix} x_d \\ x_q \end{bmatrix} = \begin{bmatrix} \cos \omega t & \sin \omega t \\ -\sin \omega t & \cos \omega t \end{bmatrix} \begin{bmatrix} x_\alpha \\ x_\beta \end{bmatrix} \qquad (5.5)$$

in which x_d *and* x_q are the variables on the *dq*-reference frame. By employing (5.5), the two DC quantities on the *dq*-reference frame can be obtained to have a better control design, especially for PI controllers.

Referring to the single-phase systems, the *αβ*-reference frame (i.e., a fictitious component x_β in-quadrature with the original variable x_α) can be created by an orthogonal signal generator (OSG), which has been generally reviewed in [32]. Then, the Park transformation in (5.5) can be employed as the same as three-phase systems.

5.3.2   Grid-connected current control

As shown in Figure 5.6, the inner grid-current loop should ensure a fast transient response, where the timescale should be far smaller than the outer voltage loop. In addition, the current controller also needs good performances in terms of power quality and grid synchronization [25]. For a single-phase inverter, periodic controllers are good candidates for zero-error tracking, e.g., the proportional resonant controller, the repetitive controller, and the deadbeat controller [33–35]. By contrast, the three-phase inverter commonly adopts a classical PI controller, which tracks the grid current reference in the synchronous *dq*-reference frame [36].

To unify and simplify the controller design for both the single-phase and three-phase inverters in this chapter, the traditional PI controller in the *dq*-reference frame will be discussed for the inner grid-current control, as presented in Figure 5.6. The *d*- and *q*-components can be calculated directly by the PQ control of the outer voltage control loop. Assuming that the controller sampling rate is fast enough and the impact from discretization can be neglected, the analysis is based on the models in the s-domain. Accordingly, this section mainly discusses the design procedure of the inner current loop for solar inverters, and the design of the DC-link voltage control and PQ control will be followed.

A. Modeling of the grid current controller

Figure 5.7 shows the circuit diagrams of the single-phase and three-phase inverters, which can be modeled as current sources according to the Kirchhoff's law. Referring to the single-phase inverter in Figure 5.7a, the dynamics for the inverter output can be described as

$$L\frac{di_g}{dt} + Ri_g = \upsilon_{AB} - \upsilon_g \tag{5.6}$$

in which L and R are the L-type inductance and its equivalent resistance, v_{AB}, i_g, and v_g are the inverter output voltage, the grid current, and the grid voltage, and A and B are the output terminals of the single-phase inverter.

Correspondingly, the dynamic equation of the three-phase inverter can be expressed as

$$
\begin{cases}
L\dfrac{di_{ga}}{dt} + Ri_{ga} = v_{AB} - v_{ga} \\[2mm]
L\dfrac{di_{gb}}{dt} + Ri_{gb} = v_{BC} - v_{gb} \\[2mm]
L\dfrac{di_{gc}}{dt} + Ri_{gc} = v_{CA} - v_{gc}
\end{cases}
\tag{5.7}
$$

in which v_{AB}, v_{BC}, v_{CA}, i_{ga}, i_{gb}, i_{gc}, and v_{ga}, v_{gb}, v_{gc} are the output voltage, the grid current, and the grid voltage of the three-phase inverter, and A, B, and C are also the output terminals. As previously mentioned in Section 5.3.1, the single-phase inverter can obtain the *dq*-reference frame by an OSG and the Park transformation in (5.5), while the three-phase inverter can do so by the reference frame transformation in (5.4) and (5.5). Therefore, it can obtain the dynamics in the *dq*-reference frame as

$$
\begin{cases}
L\dfrac{di_d}{dt} + Ri_d - \omega Li_q = v_{d1} - v_d \\[2mm]
L\dfrac{di_q}{dt} + Ri_q + \omega Li_d = v_{q1} - v_q
\end{cases}
\tag{5.8}
$$

where i_d and i_q are the grid current on the *dq*-reference frame, v_{d1} and v_{q1} are the inverter output voltages on the *d*- and *q*-axis, and v_d and v_q represent the grid voltages on the *d*- and *q*-axis, respectively.

It can be seen from (5.8) that the grid current can be controlled by regulating the inverter output voltage, where the *d*- and *q*-axis output currents are coupled to each other. Thus, the controller should add a decoupling term to obtain the output voltage references (i.e., v_{d1}^* and v_{q1}^*), which can be modified as

$$
\begin{cases}
v_{d1}^* = v_{d1} + \omega Li_q - v_d \\[2mm]
v_{q1}^* = v_{q1} - \omega Li_d - v_q
\end{cases}
\tag{5.9}
$$

The current control loop model can then be rewritten as

$$
\begin{cases}
L\dfrac{di_d}{dt} + Ri_d = v_{d1}^* \\[2mm]
L\dfrac{di_q}{dt} + Ri_q = v_{q1}^*
\end{cases}
\tag{5.10}
$$

where the *d*- and *q*-axis output currents (i.e., the grid current in grid-connected solar systems) are decoupled. In addition, the dynamics of *d*- and *q*-axis output current are identical, as demonstrated in (5.10). Consequently, the control of one axis can be analyzed in detail during design. According to (5.10), the plant transfer functions from the inverter voltage references to the grid currents in the *dq*-reference frame can be obtained as [37]

$$
\frac{i_d(s)}{v_{d1}^*(s)} = \frac{i_q(s)}{v_{q1}^*(s)} = \frac{1}{Ls + R}
\tag{5.11}
$$

also indicating that the *d*- and *q*-axis output currents have the same dynamics.

B. Design of the grid current controller

As shown in (5.10), the current control loops for the *d*- and *q*-axis components are identical. Hence, the design procedure on the *d*-axis (i.e., for the *d*-axis current i_d)

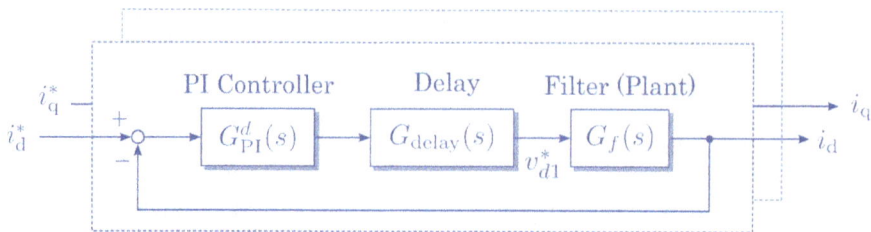

Figure 5.8   *Current control loops in the synchronous* dq-*reference frame, which focus on the design of the controllers:* $G_{PI}^{d}(s)$ = *the* d-*component PI current controller,* $G_{delay}(s)$ = *the elapsed delay due to the PWM and computations in the control system, and* $G_f(s)$ = *the filter (plant) transfer function*

is depicted in Figure 5.8, which is also applicable in the controller design for the q-axis current.

As presented in Figure 5.8, these transfer functions of $G_{PI}^{d}(s)$, $G_{delay}(s)$, and $G_f(s)$ can be given as

$$G_{PI}^{d}(s) = k_{dp} + \frac{k_{di}}{s} = \frac{k_{dp}(1 + T_{di}s)}{T_{di}s} \tag{5.12}$$

$$G_{delay}(s) = \frac{1}{1 + 1.5T_s s} \tag{5.13}$$

$$G_f(s) = \frac{1}{R + Ls} = \frac{T_f}{L(1 + T_f s)} \tag{5.14}$$

where k_{dp} and k_{di} are the proportional and the integral gains of the PI current controller, $T_{di} = k_{dp}/k_{di}$ is the integrator time constant, T_s is the sampling time, and $T_f = L/R$ is the L-type filter time constant.

The cross-coupling ($\omega L i_q$) and the voltage feed-forward (v_d) terms in (5.9) for the d-axis current component control loop are neglected, which is illustrated in Figure 5.8. Therefore, the two terms are considered as disturbances in the current control system. Thus, the open-loop transfer function can be expressed as

$$G_{open}^{d}(s) = G_{PI}^{d}(s) G_{delay}(s) G_f(s) = \frac{k_{dp}T_f(1 + T_{di}s)}{T_{di}Ls(1 + T_f s)(1 + 1.5T_s s)} \tag{5.15}$$

Accordingly, the closed-loop transfer function is obtained as

$$G_{close}^{d}(s) = \frac{G_{open}^{d}(s)}{1 + G_{open}^{d}(s)} = \frac{k_{dp}T_f(1 + T_{di}s)}{T_{di}Ls(1 + T_f s)(1 + 1.5T_s s) + k_{dp}T_f(1 + T_{di}s)} \tag{5.16}$$

which can also be applied for the q-axis current.

To simplify the parameters design of the PI current controller, the integrator time T_{di} is chosen as the same as the filter time constant T_f in (5.16). In that case, the closed-loop transfer function can be rewritten as

$$G_{close}^{d}(s) = \frac{k_{dp}}{Ls(1 + 1.5T_s s) + k_{dp}} = \frac{\frac{2k_{dp}}{3T_s L}}{s^2 + \frac{2}{3T_s}s + \frac{2k_{dp}}{3T_s L}} \tag{5.17}$$

being a typical second-order system with

$$\omega_n^2 = \frac{2k_{dp}}{3T_s L} \text{ and } 2\zeta\omega_n = \frac{2}{3T_s} \tag{5.18}$$

where ω_n is the natural frequency and ζ is the damping ratio.

In practice, the optimal damping ratio is $\zeta = 1/\sqrt{2}$ to achieve an overshoot of 5 percent for a step response [33]. Consequently, the proportional and integral gains can be obtained as

$$k_{dp} = \frac{L}{3T_s} \text{ and } k_{di} = \frac{L}{3T_s T_f} \tag{5.19}$$

Assuming that the grid current controller is optimally designed, the closed-loop transfer function can then be approximated as

$$G_{close}^d (s) \approx \frac{1}{1 + 3T_s s} = \frac{1}{1 + \tau s} \tag{5.20}$$

in which the bandwidth can be estimated as

$$f_{bw}^d \approx \frac{1}{2\pi\tau} = \frac{1}{6\pi T_s} \tag{5.21}$$

It should be noted that the reference frame transformations will increase calculational burdens in practice. Therefore, the current controller has another alternative method to achieve zero-error tracking, i.e., adopting a PR controller in the single-phase inverters or the three-phase inverters under the $\alpha\beta$-reference frame [33, 34] (i.e., both are typical periodic signal systems). The PR controller has better performance in terms of zero-error tracking than the PI controller in AC signal systems, where the transfer function can be expressed as

$$G_{PR} (s) = k_{rp} + \frac{k_{ri}s}{s^2 + \omega^2} \tag{5.22}$$

where k_{rp} and k_{ri} represent the control gains of the PR controller $G_{PR}(s)$.

It can be illustrated in (5.22) that the PR controller can approach infinity at its resonant frequency ω, resulting in good tracking for the AC signal. Referring to parameters design for the PR controller, k_{rp} can be tuned in the same way in (5.19), whereas k_{ri} can be chosen as

$$k_{ri} = 2\alpha_h k_{rp} \tag{5.23}$$

where α_h is the resonant bandwidth. To maintain control stability, the resonant bandwidth should be far smaller than the current controller bandwidth.

Accordingly, the PR controller has a significant advantage in AC periodic-signal systems, reducing the complexity caused by the reference frame transformation. Obviously, a small variation of the resonant frequency ω will affect the PR controller's performance in (5.22), possibly resulting in system instability. Therefore, an improved PR controller is introduced in [38] to have an adjustable tracking frequency, where the transfer function can be expressed as

$$G_{PR} (s) = k_{rp} + \frac{k_{ri}\omega_c s}{s^2 + 2\omega_c s + \omega^2} \tag{5.24}$$

in which ω_c represents the adjustable cut-off frequency. Notably, the cut-off frequency ω_c will lead to a compromise of the controller gain [37–39].

5.3.3   PQ Control

A. Modeling of the PQ control

As mentioned above, the advanced grid-connected solar system should have flexible power control capability, i.e., the full controllability of the active power and reactive power [2]. The three-phase *PQ* control is based on the *instantaneous power theory* [40]. In the synchronous *dq*-reference frame, the instantaneous active power *P* and reactive power *Q* can be given as

$$\begin{cases} P = \dfrac{3}{2}\left(v_d i_d + v_q i_q\right) \\ Q = \dfrac{3}{2}\left(v_q i_d - v_d i_q\right) \end{cases} \tag{5.25}$$

Assuming that the PLL is aligned with the grid voltage vector to the *d*-axis of the *dq*-reference frame (i.e., $v_q = 0$), the transfer functions from the *d*- and *q*-axis output currents to the active and reactive power can be calculated as

$$\begin{aligned} \frac{P(s)}{i_d(s)} &= \frac{3}{2} v_d(s) = \frac{3}{2} V_m \\ \frac{Q(s)}{i_q(s)} &= -\frac{3}{2} v_d(s) = -\frac{3}{2} V_m \end{aligned} \tag{5.26}$$

which are simply the proportional gains with V_m being the amplitude of the grid voltage v_g. Notably, for single-phase inverters, the power calculation and the single-phase PQ theory can be achieved similarly.

B. Design of the DC-link controller

Figure 5.9 shows the PQ control block, which includes the DC-link voltage control in the *dq*-reference frame. According to (5.25) and (5.26), an open-loop control is sufficient to regulate the active and reactive power with the direct power references. That is, the active power and reactive power references (i.e., P^* and Q^*) can be employed directly to obtain the *d*- and *q*-axis currents when the DC-link voltage

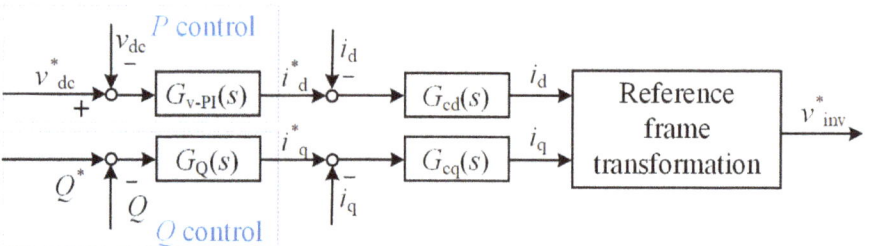

Figure 5.9   *PQ control block with DC-link voltage control loop in the synchronous* dq-*reference frame, where* G_Q(s) *represents the reactive power controller,* G_{cd}(s) *and* G_{cq}(s) *are the current controllers for the* d- *and* q-*axis components, and* v*$_{inv}$ *is the reference inverter output voltage*

is an ideal DC source. However, a closed-loop control is practically employed on the active power control to improve the control performance due to uncertainties in the system (e.g., input power variations and power losses), as demonstrated in Figure 5.9 (i.e., the active power is regulated by a closed-loop DC-link voltage controller). The DC-link voltage controller design will be detailed in Section 5.3.4.

5.3.4 *DC-link voltage control*

The main objective of the outer control loop of the DC-link voltage can be summarized in two aspects: (1) alleviating the pulsating power effects on the inverter (i.e., the double-frequency ripple of the single-phase inverter and the six-frequency ripple of the three-phase inverter); (2) minimizing the fluctuations caused by the transient input power change from the PV array [12]. The most frequently used DC-link voltage controllers are based on a standard PI controller [41]. In this section, the basic PI controller design for the DC-link voltage will be introduced.

A. Modeling of the DC-link controller

Assuming that the power in the DC side is transferred to the AC side without losses, the input DC power is equal to the inverter output average power based on the instantaneous power theory, which can be given as

$$\overbrace{v_{dc}\left(i_{dc}+C_{dc}\frac{dv_{dc}}{dt}\right)}^{\text{DC input power}} = \overbrace{\frac{1}{2}\left(v_d i_d+v_q i_q\right)\text{(single-phase)}\ or\ \frac{3}{2}\left(v_d i_d+v_q i_q\right)\text{(three-phase)}}^{\text{Output average power}} \tag{5.27}$$

where C_{dc} is the DC-link capacitance as shown in Figure 5.6. Here, the *d*-axis reference of the solar inverter is synchronized with the *d*-axis component of the grid voltage by a PLL, which means $v_q = 0$. In that case, the linear model can be achieved by applying the small-signal analysis to (5.27). In that case, the transfer functions of the DC-link voltage in the single-phase inverter and the three-phase inverter can be expressed as

$$\frac{v_{dc}(s)}{i_g(s)} = \frac{1}{2}\frac{V_m}{V_{dc}C_{dc}s}\ \text{(single-phase)} \tag{5.28}$$

$$\frac{v_{dc}(s)}{i_d(s)} = \frac{3}{2}\frac{V_m}{V_{dc}C_{dc}s}\ \text{(three-phase)} \tag{5.29}$$

in which V_{dc} and I_{dc} are the average DC-link voltage (v_{dc}) and current (i_{dc}), respectively.

B. Design of the DC-link controller

As presented in Figure 5.9, the outer DC-link voltage controller can generate the *d*-axis current reference. In addition, the traditional and effective PI controller for the DC-link voltage will be introduced, where the transfer functions can be expressed as

$$G_v(s) = k_{vp}+\frac{k_{vi}}{s} = \frac{k_{vp}(1+T_{vi}s)}{T_{vi}s} \tag{5.30}$$

$$G_v(s) = \frac{1}{2}\frac{V_m}{V_{dc}C_{dc}s}\ \text{(single-phase)}\ or\ \frac{3}{2}\frac{V_m}{V_{dc}C_{dc}s}\ \text{(three-phase)} \tag{5.31}$$

where k_{vp} and k_{vi} represent the proportional and integral parameters of the PI controller for the DC-link voltage, respectively, and $T_{vi} = k_{vp}/k_{vi}$ is the integrator time.

Referring to the choice of the DC-link voltage reference, two aspects should be considered. One is that the DC-link voltage V_{dc} should be higher than the minimum required value in different applications and control conditions to ensure the current controllability and avoid the overmodulation of grid-connected solar inverters. That can be summarized as follows:

- In single-phase systems, $V_{dc} \geq V_m$.
- In three-phase systems, $V_{dc} \geq \sqrt{3} V_m$ with space vector modulation scheme, while $V_{dc} \geq 2 V_m$ with a sinusoidal PWM scheme.

The other aspect is that the average DC-link voltage V_{dc} should not be much higher than the above required value to ensure low power losses (i.e., high DC-link voltage leads to high switching losses for the semiconductor devices).

According to Figures 5.6, 5.8 and 5.9, the open-loop transfer function of the DC-link voltage control loop can be expressed by

$$G_{v\text{-open}}(s) = G_{v\text{-PI}}(s) \, G_{close}^d(s) \, G_P(s) \tag{5.32}$$

With the simplified current control loop in (5.20), the transfer function in (5.32) is given as

$$G_{v\text{-open}}(s) \approx \frac{3 V_m k_{vp}(1 + T_{vi}s)}{2 T_{vi} V_{dc} C_{dc} s^2 (1 + 3 T_s s)} \tag{5.33}$$

Then, the phase crossover frequency ω_{pc} can be expressed as

$$\omega_{pc}(s) = \frac{1}{\sqrt{3 T_{vi} T_s}} \tag{5.34}$$

Accordingly, the parameters of the PI controller k_{vp} and k_{vi} at the phase crossover frequency (ω_{pc}) is given by

$$k_{vp} = \frac{C_{dc}}{2\sqrt{T_{vi} T_s}}, \text{ and } k_{vi} = \frac{C_{dc}}{2\sqrt{T_{vi}^3 T_s}} \tag{5.35}$$

The parameters can be selected with the consideration of the desired bandwidth or the response time, the phase margin or the transient behavior of the solar inverter system.

5.4 Case study

5.4.1 PI controller for three-phase inverters

To demonstrate the control design in Section 5.3, the controllers of a 10-kW two-stage three-phase grid-connected solar inverter are designed. Figure 5.10 shows the schematic and the entire control of the inverter stage, which includes the reference frame transformations, the current control in the *dq*-reference frame, the PQ control, and the DC-link voltage control. The system parameters of the two-stage

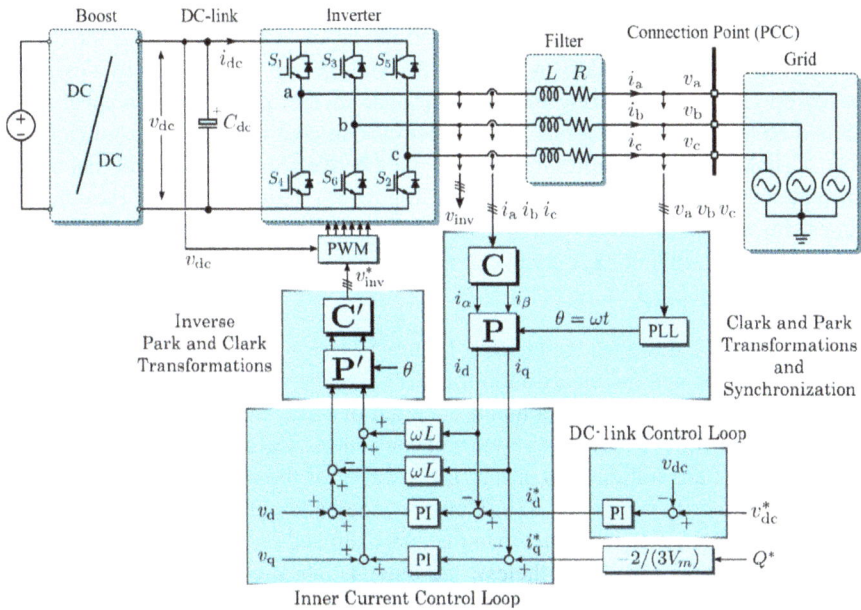

Figure 5.10 *Control structure of the inverter stage in a two-stage, three-phase grid-connected solar system in the synchronous* dq-*reference frame, where the PQ control includes a closed-loop voltage control and an open-loop control for the reactive power. The PLL is used for reference transformations.*

three-phase solar system are shown in Table 5.1. Assuming that the MPPT control in the first stage is robust, the input power for the three-phase inverter is constant.

Based on (5.19) and the system parameters in Table 5.1, the PI control parameters for the *d*-axis current loop can be chosen as

$$k_{dp} = 33.3 \text{ and } k_{di} = 666.7 \tag{5.36}$$

Table 5.1 *System parameters of the 10-kW grid-connected three-phase solar inverter*

Parameter	Symbol	Value
DC-link voltage reference	v^*_{dc}	600 V
DC-link capacitor	C_{dc}	500 μF
Grid phase voltage amplitude	V_m	311 V
Filter inductance	L	5 mH
Filter resistance	R	0.1 Ω
Switching frequency	f_{sw}	20 kHz
Sampling frequency	$f_{sw} = 1/T_s$	20 kHz

where the bandwidth can be approximated as $f_{bw}^d = 1$ kHz according to (5.21). Otherwise, the PI controller for the DC-link voltage can be analytically designed based on (5.35). Considering that the timescale of the outer voltage control loop should be much larger than that of the inner current control loop, the bandwidth of the DC-link voltage controller should be limited between 1/50 and 1/10 of that in the current control loop [42]. Consequently, a bandwidth frequency of 100 Hz (i.e., ten times slower than the inner current control loop) is selected, and then the PI parameters for the DC-link voltage controller are calculated as

$$k_{vp} = 0.27 \text{ and } k_{vi} = 16.11 \tag{5.37}$$

According to the designed parameters, Figure 5.11 shows the frequency responses of the open-loop and the closed-loop transfer functions of the two control loops. It can be seen in Figure 5.11a that a phase margin of 78.8° and 65.5° for the DC-link voltage PI controller and the d-axis current controller, respectively, is achieved. The phase margins are sufficient to ensure the stability of the closed-loop control systems, as validated by Figure 5.11b. However, the bandwidth of the d-axis current control loop (1.45 kHz) is higher than the approximated one (1 kHz), which is also shown in Figure 5.11b. Nevertheless, the outer DC-link voltage control loop has a bandwidth of 92.5 Hz, which satisfies the timescale requirements for the double closed-loop controller. In that case, the inner current loop can have a fast dynamic tracking for the outer voltage loop.

With the above parameters, a two-stage three-phase grid-connected solar inverter system is built up in MATLAB®. Then, the designed controller parameters (i.e., (5.36) and (5.37)) are applied to the three-phase grid-connected solar inverter. Simulation results are shown in Figure 5.12, which includes the performance of the grid line-to-line voltages, the grid currents, the DC-link voltage, and the dq-current components. There are two transient operations in the simulations, i.e., one is a step change in the active power from 0 to 8 kW at $t = 0.2$ s, and the other is a step change of the reactive power from 0 to 6 kVAR at $t = 0.4$ s. Specifically, when the input power is changed at $t = 0.2$ s, the d-axis current fast-tracks the change. As shown in Figure 5.12c, the DC-link voltage appears as a 7.5 percent overshoot and recovers to steady-state after one cycle. Moreover, the q-axis current step change occurs at $t = 0.4$ s, which can be achieved by directly setting the reactive power reference according to the open-loop control in Figure 5.10. That is an effective and simple reactive power control method under an abnormal grid voltage condition (e.g., the low-voltage ride-through). Furthermore, Figure 5.12e indicates that the q-axis current can regulate quickly when there is a disturbance in the q-axis current, i.e., the step change at $t = 0.2$ s, which verified the effectiveness of the PI controller for the q-axis current.

In all, the designed controllers of the three-phase grid-connected solar inverter in Figure 5.10 can have a good dynamic performance (i.e., accuracy and quickly tracking the disturbance). Notably, the grid-connected solar inverter system is always affected by the background harmonics of the real utility grid, resulting in serious distortions on the grid current. To alleviate this, a harmonic compensation should

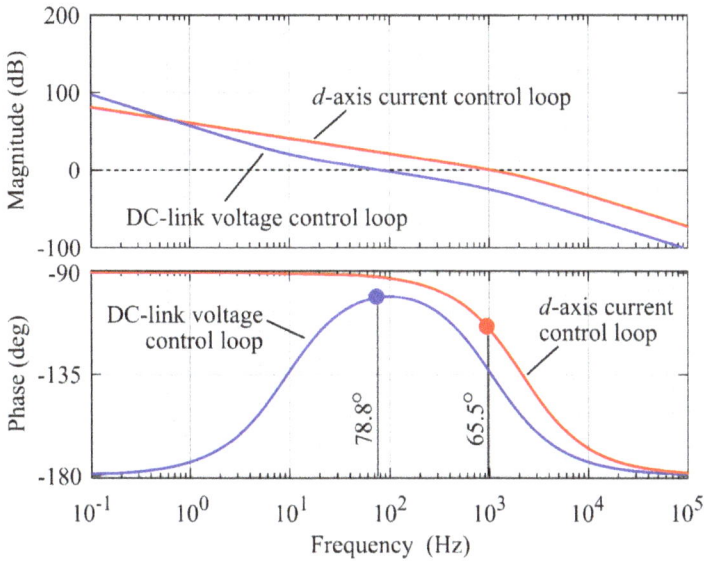

(a)

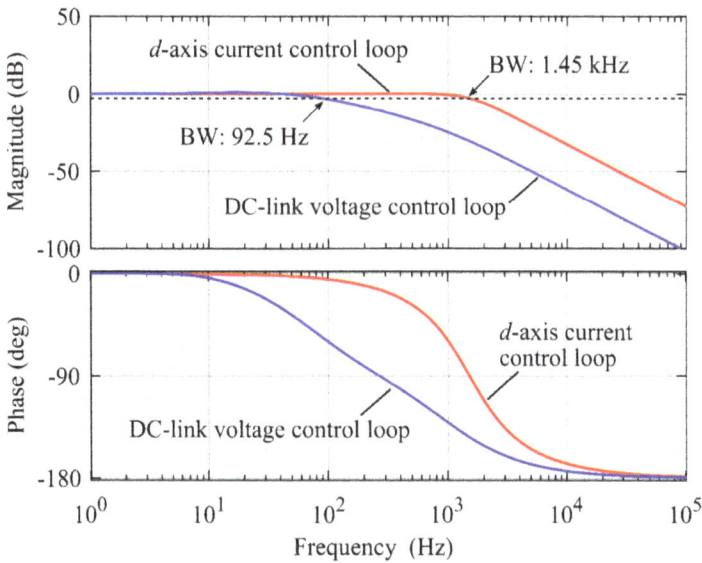

(b)

Figure 5.11 *Frequency response of the double-loop control (i.e., the DC-link voltage control loop and the* d-*axis current control loops) with the designed parameters: (a) open loop Bode plots and (b) closed-loop Bode plots, where BW represents bandwidth of the system*

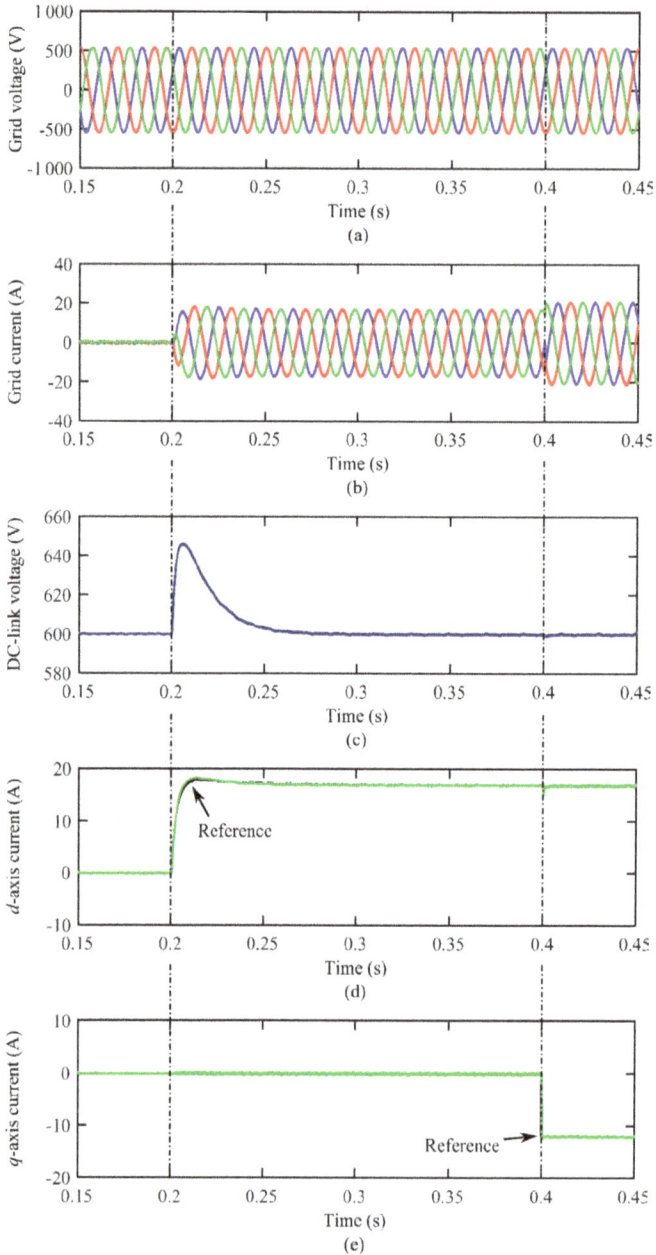

Figure 5.12   *Simulation results for the grid-connected three-phase AC–DC*
converter system controlled in the synchronous reference frame
with the designed parameters: (a) grid line-to-line voltages, (b)
grid currents, (c) DC-link voltage, (d) d-*axis current component,*
and (e) q-*axis current*

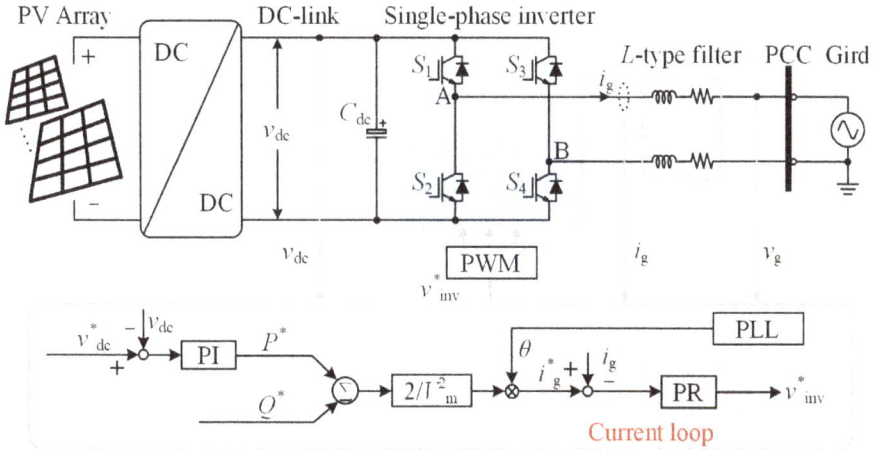

Figure 5.13   *Control structure of the single-phase grid-connected solar inverter, where the current loop adopts PR controller*

be integrated [33]. The case study in this chapter aims to provide an exercise for the audience to better understand the control design procedures of grid-connected solar inverters.

5.4.2 PR controller for single-phase inverters

The PI controller in the *dq*-reference frame can be employed in both the three-phase and single-phase inverters, and yet the reference frame transformation for single-phase inverters will increase the calculation complexity. Alternatively, there are other advanced solutions to have zero-error tracking without the reference frame transformation, e.g., the PR controllers and repetitive controllers, which have good tracking performance in AC periodical signal systems [43]. In this case, the control loop can still be taken as a cascaded-control system, which includes an outer voltage or power control and an inner current control. Generally, the design of the period control can also be applied directly to the single-phase inverter or in the *αβ*-reference frame of the three-phase inverter. When considering without the effort of the Park transformation (*αβ→dq*), this control method, thus, is more beneficial to single-phase inverters.

Accordingly, an example of the current controller implemented with the PR controller in a single-phase inverter is demonstrated in Figure 5.13, being without the Park transformation (*αβ→ dq*). Besides, the system parameters of a 3.5-kW grid-connected single-phase inverter are given in Table 5.2, while the controller parameters are designed according to the discussion in Subsection 5.3.2. Through (5.24), the controller parameters can be calculated as

$$k_{\mathrm{rp}} = 33.3 \text{ and } k_{\mathrm{ri}} = 13\ 334 \tag{5.38}$$

Table 5.2   System parameters of the 3.5-kW grid-connected single-phase inverter.

Parameter	Symbol	Value
DC-link voltage reference	v_{dc}^*	400 V
DC-link capacitor	C_{dc}	1 000 μF
Grid phase voltage amplitude	V_m	311 V
Filter inductance	L	7.6 mH
Filter resistance	R	0.08 Ω
Switching frequency	f_{sw}	10 kHz
Sampling frequency	$f_{sw} = 1/T_s$	10 kHz

Figure 5.14 presents the simulation performance of the single-phase inverter with a current step change from 5 A to 10 A at $t = 0.205$ s. To compare with the PI control, the dynamic performance of the single-phase inverter has been transformed in the dq-reference frame, as shown in Figure 5.14. The results in Figure 5.14 indicate that the grid current can track the reference quickly and accurately, verifying the controllability of the PR controller. However, when compared to the simulation performance of the PI controller, the dynamics are slightly different in terms of overshoots and settling time, which may be related to the coupling effects and the reference frame in the three-phase solar inverter. Nevertheless, this case study with the PR controller illustrates that the control of single-phase solar inverters can be implemented in both the $\alpha\beta$-reference frame and the dq-reference frame. In addition, the PR controller can also be employed in the $\alpha\beta$-reference frame of the three-phase solar inverter.

In all, the two case studies, i.e., the PI controller in the dq-reference frame for the three-phase inverter and the PR controller in the $\alpha\beta$-reference frame for single-phase

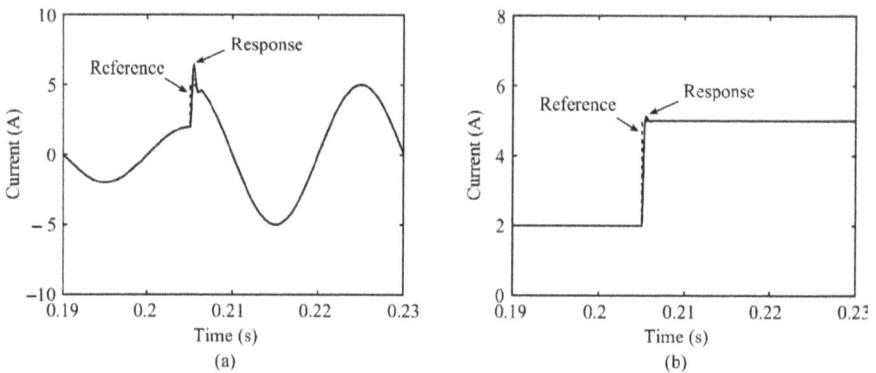

Figure 5.14   Dynamic performance of the single-phase inverter with the PR controller: (a) the grid current (i.e., the α-axis current) and (b) the d-axis current.

inverter, provide full control design procedures, which should consider the performance in terms of dynamic response (i.e., accuracy and tracking speed) and steady-state characteristics.

5.5   Summary

In this chapter, the general control for grid-connected solar inverters has been introduced, which includes the MPPT algorithm, the PQ control (e.g., the DC-link voltage control), and the grid current control. The brief introduction for the modeling and MPPT control of the PV panels is first presented, where the traditional P&O MPPT algorithm is exemplified to demonstrate the PV characteristics and the MPPT function. Then, the control procedures of grid-connected solar inverters (i.e., single-phase and three-phase inverters) have been exhibited, mainly designed in the *dq*-reference frame. The common procedures contain the reference frame transformations, the modeling and design of the controller, and the parameters tuning. Except for the PI controller in the *dq*-reference frame for the three-phase inverter, the design of a typical PR controller for a single-phase inverter (i.e., exemplified as an AC period system) has also been demonstrated. Finally, simulations are carried out to validate the controllability of the basic control strategies for grid-connected solar inverters.

References

[1]   Blaabjerg F., Yang Y., Yang D., Wang X. 'Distributed power-generation systems and protection'. *Proceedings of the IEEE*. 2017;**105**(7):1311–31.

[2]   IEEE Standard Committee. 'IEEE standard for interconnection and interoperability of distributed energy resources with associated electric power systems interfaces'. *IEEE Std 1547-2018*. 2018:1–138.

[3]   Blaabjerg F., Ionel D.M. *Renewable Energy Devices and Systems with Simulations in MATLAB® and ANSYS®*. 1st edn. CRC Press; 2017.

[4]   Wu Y.K., Lin J.H., Lin H.J. 'Standards and guidelines for grid-connected photovoltaic generation systems: a review and comparison'. *IEEE Transactions on Industry Applications*. 2017;**53**(4):3205–16.

[5]   Verband der Elektrotechnik. *Power Generation Systems Connected to the Low-Voltage Distribution Network–Technical Minimum Requirements for the Connection to and Parallel Operation with Low-Voltage Distribution Networks (VDE-AR-N 4105)*. Frankfurt, Germany: Verband der Elektrotechnik Aug; 2011.

[6]   Podder A.K., Roy N.K., Pota H.R. 'MPPT methods for solar PV systems: a critical review based on tracking nature'. *IET Renewable Power Generation*. 2020;**14**(9):1752–1424.

[7]   Yang Y., Wang H., Blaabjerg F. 'Reactive power injection strategies for single-phase photovoltaic systems considering grid requirements'. *IEEE Transactions on Industry Applications*. 2014;**50**(6):4065–76.

[8]   Sangwongwanich A., Yang Y., Blaabjerg F. 'A sensorless power reserve control strategy for two-stage grid-connected PV systems'. *IEEE Transactions on Power Electronics*. 2017;**32**(11):8559–69.

[9]   Rakhshani E., Rodriguez P. 'Inertia emulation in AC/DC interconnected power systems using derivative technique considering frequency measurement effects'. *IEEE Transactions on Power Systems*. 2017;**32**(5):3338–51.

[10] Fang J., Tang Y., Li H., Li X. 'A battery/ultracapacitor hybrid energy storage system for implementing the power management of virtual synchronous generators'. *IEEE Transactions on Power Electronics*. 2018;**33**(4):2820–4.

[11] Andresen M., Ma K., Buticchi G., Falck J., Blaabjerg F., Liserre M. 'Junction temperature control for more reliable power electronics'. *IEEE Transactions on Power Electronics*. 2018;**33**(1):765–76.

[12] Blaabjerg F., Teodorescu R., Liserre M., Timbus A.V. 'Overview of control and grid synchronization for distributed power generation systems'. *IEEE Transactions on Industrial Electronics*. 2006;**53**(5):1398–409.

[13] Villalva M.G., Gazoli J.R., Filho E.R. 'Comprehensive approach to modeling and simulation of photovoltaic arrays'. *IEEE Transactions on Power Electronics*. 2009;**24**(5):1198–208.

[14] Esram T., Chapman P.L. 'Comparison of photovoltaic array maximum power point tracking techniques'. *IEEE Transactions on Energy Conversion*. 2007;**22**(2):439–49.

[15] Yang Y., Blaabjerg F. 'A modified P&O MPPT algorithm for single-phase PV systems based on deadbeat control'. *Proc. of IET PEMD'12*. 2012:1–5.

[16] Rezk H., Fathy A., Abdelaziz A.Y. 'A comparison of different global MPPT techniques based on meta-heuristic algorithms for photovoltaic system subjected to partial shading conditions'. *Renewable and Sustainable Energy Reviews*. 2017;**74**(6):377–86.

[17] Ishaque K., Salam Z., Amjad M., Mekhilef S. 'An improved particle swarm optimization (PSO)–based MPPT for PV with reduced steady-state oscillation'. *IEEE Transactions on Power Electronics*. 2012;**27**(8):3627–38.

[18] Ahmed J., Salam Z. 'A maximum power point tracking (MPPT) for PV system using cuckoo search with partial shading capability'. *Applied Energy*. 2014;**119**:118–30.

[19] Blaabjerg F., Chen Z., Kjaer S.B. 'Power electronics as efficient interface in dispersed power generation systems'. *IEEE Transactions on Power Electronics*. 2004;**19**(5):1184–94.

[20] Teodorescu R., Liserre M., Rodriguez P. *Grid Converters for Photovoltaic and Wind Power Systems*. New York, NY, USA: Wiley; 2011.

[21] Romero-Cadaval E., Francois B., Malinowski M., Zhong Q.-C. 'Grid-connected photovoltaic plants: an alternative energy source, replacing conventional sources'. *IEEE Industrial Electronics Magazine*. 2015;**9**(1):18–32.

[22] Yang Y., Enjeti P., Blaabjerg F., Wang H. 'Wide-scale adoption of photovoltaic energy: grid code modifications are explored in the distribution grid'. *IEEE Industry Applications Magazine*. 2015;**21**(5):21–31.

[23] Vasquez J.C., Mastromauro R.A., Guerrero J.M., Liserre M. 'Voltage support provided by a droop-controlled multifunctional inverter'. *IEEE Transactions on Industrial Electronics*. 2009;**56**(11):4510–19.

[24] Cagnano A., De Tuglie E., Liserre M., Mastromauro R.A. 'Online optimal reactive power control strategy of PV inverters'. *IEEE Transactions on Industrial Electronics*. 2011;**58**(10):4549–58.

[25] Zhong Q.-C., Hornik T. 'Cascaded current–voltage control to improve the power quality for a grid-connected Inverter with a local load'. *IEEE Transactions on Industrial Electronics*. 2013;**60**(4):1344–55.

[26] Carnieletto R., Brandão D.I., Farret F.A., Simões M.G., Suryanarayanan S. 'Smart grid initiative: a multifunctional single-phase voltage source inverter'. *IEEE Industry Applications Magazine*. 2011;**17**(5):27–35.

[27] Yang Y., Blaabjerg F., Wang H. 'Low-voltage ride-through of single-phase transformerless photovoltaic inverters'. *IEEE Transactions on Industry Applications*. 2014;**50**(3):1942–52.

[28] O'Rourke C.J., Qasim M.M., Overlin M.R., Kirtley J.L., C. J. O'Rourke, M. M., Qasim M.R.O. 'A geometric interpretation of reference frames and transformations: dq0, Clarke, and Park'. *IEEE Transactions on Energy Conversion*. 2019;**34**(4):2070–83.

[29] Clarke E. 'Circuit analysis of AC power systems; symmetrical and related components. Vol. 1'. Wiley; 1943.

[30] Chen D., Zhang J., Qian Z. 'An improved repetitive control scheme for grid-connected inverter with frequency-adaptive capability'. *IEEE Transactions on Industrial Electronics*. 2013;**60**(2):814–23.

[31] Park R.H. 'Two-reaction theory of synchronous machines generalized method of analysis - Part I'. *Transactions of the American Institute of Electrical Engineers*. 1929;**48**(3):716–27.

[32] Teodorescu R., Liserre M., Rodriguez P. *Grid Converters for Photovoltaic and Wind Power Systems*. New York, NY, USA: Wiley; 2011.

[33] Teodorescu R., Blaabjerg F., Liserre M., Loh P.C. 'Proportional-resonant controllers and filters for grid-connected voltage - source converters'. *IEE Proceedings - Electric Power Applications*. 2006;**153**(5):750–62.

[34] Yang Y., Zhou K., Wang H., Blaabjerg F., Wang D., Zhang B. 'Frequency adaptive selective harmonic control for grid-connected Inverters'. *IEEE Transactions on Power Electronics*. 2015;**30**(7):3912–24.

[35] Stumper J.-F., Hagenmeyer V., Kuehl S., Kennel R. 'Deadbeat control for electrical drives: a robust and performant design based on differential flatness'. *IEEE Transactions on Power Electronics*. 2015;**30**(8):4585–96.

[36] Abdelhakim A., Mattavelli P., Boscaino V., Lullo G. 'Decoupled control scheme of grid-connected split-source inverters'. *IEEE Transactions on Industrial Electronics*. 2017;**64**(8):6202–11.

[37] Liserre M., Teodorescu R., Blaabjerg F. 'Multiple harmonics control for three-phase grid converter systems with the use of PI-RES current controller in a rotating frame'. *IEEE Transactions on Power Electronics*. 2006;**21**(3):836–41.

[38] Tan P.C., Loh P.C., Holmes D.G. 'High-performance harmonic extraction algorithm for a 25 kV traction power quality conditioner'. *IEE Proceedings - Electric Power Applications*. 2004;**151**(5):505–12.

[39] Pan D., Ruan X., Bao C., Li W., Wang X. 'Optimized controller design for LCL-type grid-connected inverter to achieve high robustness against grid-impedance variation,'. *IEEE Trans. Ind. Electron*. 2015;**62**(3):1537–47.

[40] Akagi H., Kanazawa Y., Nabae A. 'Instantaneous reactive power compensators comprising switching devices without energy storage components'. *IEEE Transactions on Industry Applications*. 1984;**IA-20**(3):625–30.

[41] Merai M., Naouar M.W., Slama-Belkhodja I., Monmasson E. 'An adaptive Pi controller design for DC-link voltage control of single-phase grid-connected converters'. *IEEE Transactions on Industrial Electronics*. 2019;**66**(8):6241–9.

[42] Zhou D., Blaabjerg F. 'Bandwidth oriented proportional-integral controller design for back-to-back power converters in DFIG wind turbine system'. *IET Renewable Power Generation*. 2017;**11**(7):941–51.

[43] Zhou K., Lu W., Yang Y., Blaabjerg F. 'Harmonic control: a natural way to bridge resonant control and repetitive control'. *Proceedings of the American Control Conference*. 2013:3189–93.

Chapter 6

Control strategy for grid-connected solar inverter for IEC standards

Mohammed Ali Khan[1] and Ariya Sangwongwanich[2]

The rapid evolution of renewable-based power generation has increased the integration of distributed generation (DG) units recently. Increase in power capacity of the DGs and its feeding to the utilities has triggered many concerns. A vast range of fault, i.e., destabilization of grid due to unregulated power injection, can cause a complete grid collapse. Hence, to tackle such issues and create a secure perimeter for grid operation, utility companies along with the governments of different countries revised the grid codes. These grid codes ensure that the fault, such as frequency mismatch, overvoltage, and undervoltage is detected and depending upon the severity of the fault, appropriate action is performed by controlling the inverter to stabilize the anomaly. If the fault tends to persist even after a prescribed fault tolerance limit, the grid is disconnected from the DGs to prevent any further harm to the utilities as well as the DGs. All the important parameters such as operating voltage with fluctuation limits, frequency fluctuation limit, and permissible time up to which the fluctuation can be tolerated, are prescribed by the grid codes. The grid code enables the low-voltage ride-through (LVRT) capability of the DGs by providing set of operating instruction. In this chapter, a comparative analysis between different grid codes focusing on LVRT requirement and islanding criteria is presented along with the analysis of different control techniques proposed for islanding detection and reactive power injection.

6.1 LVRT requirement for control

The grid-connected photovoltaic (PV) system (GCPVS) is considered to be in LVRT condition when a certain criterion in the grid code is not maintained [1]. Hence, it is necessary to understand grid code and its focus on different aspects of grid integration. Various parameters such as operating frequency, operating voltage, power

[1]Advance Power Electronics and Research Laboratory, Department of Electrical Engineering, Jamia Millia Islamia (A Central University), New Delhi, India
[2]Center of Reliable Power Electronics (CORPE), Department of Energy Technology, Aalborg University, Aalborg, Denmark

Table 6.1   Permissible limit of voltage fluctuation with respect to nominal voltage

Country	Grid code	Requirement	
		Rating	Limit range (%)
United States	Federal Energy Regulatory Commission (2005) [5]	–	±10
United Kingdom	National Grid (2010) The grid code [6]	132 kV and 275 kV 400 kV	±10 ±5
Ireland	EirGrid [7] EirGrid grid code [7]	110 kV 220 kV 400 kV	−10 to +12 −9 to +12 −13 to +5
China	China's National Standard: GB/T 19963—2011 [8]	>110 V ≤110 V	±8 ±10
Germany	German e-on Grid code [9]	110 kV 220 kV 400 kV	−12.7 to +11.8 −12.3 to +11.4 −7.3 to +10.5
India	CERC (Indian Electricity Grid Code) Regulations, 2010 [10]	–	±6
Denmark	Technical regulation 3.2.1 for power plants up to and including 11 kW [11]	132 kV 150 kV 400 kV	−10 to +10 −10 to +13 −10 to +5
Australia	Australia Standards Electromagnetic Compatibility (EMC) AS61000.3.100 [12]	–	+10 to −6

factor, and reactive power are monitored at the point of common coupling (PCC) [2, 3]. In this section, an overview of grid codes for different countries is presented regarding LVRT.

6.1.1   Permissible limit for voltage fluctuation

The voltage at PCC is monitored and it is maintained to ensure a continuous operation even when the voltage varies under a certain range as specified by grid code [4]. The range of voltage prescribed by different countries is listed in Table 6.1.

It is required that the GCPVS operates within the voltage fluctuation limits at PCC. The voltage control is in coordination with the reactive power. When the voltage is under the fluctuation limit at PCC it must also be ensured that the DGs are operating under the operating voltage range of the equipment present. Preferably the operating voltage range is between 0.9 p.u. and 1.1 p.u. [13].

6.1.2   Permissible limit for frequency fluctuation

Majority of the DGs utilizing synchronous generator for power production. The frequency can be regulated by controlling the rotor speed of synchronous generator [14]. If there is a

Table 6.2   *Permissible limit of frequency fluctuation with respect to nominal*
frequency

| Country | Grid code | Requirement | |
		Frequency range (Hz)	Tolerance limit
United States	Federal Energy Regulatory Commission (2005) [5]	59.95–60.05	Normal operation
United Kingdom	National Grid (2010) The grid code [6]	49–51	Normal operation
		51.5–52	15 min
		51–51.5	90 min
		47.5–49	90 min
		47–47.5	20 s
Ireland	EirGrid [7] EirGrid grid code [7]	49.5–50.5	Normal operation
		50.5–52	60 min
		49.5–47.5	60 min
		<47.5	20 s
China	China's National Standard: GB/T 19963—2011 [8]	49.8–50.2	Normal operation
		50.2–52	2 min
		48–49.8	10 min
Germany	German e-on Grid code [9]	47.5–51.5	Normal operation
India	CERC (Indian Electricity Grid Code) Regulations, 2010 [10]	48.5–51.5	Normal operation
Denmark	Technical regulation 3.2.1 for power plants up to and including 11 kW [11]	49.5–50.2	Normal operation
		49–49.5	5 h
		48–49	30 min
		50.2–52	15 min
		47.5–48	3 min
		47–47.5	20 s
Australia	Australia Standards Electromagnetic compatibility (EMC) AS61000.3.100 [12]	49.75–50.25	Normal operation

frequency mismatch between generation and power consuming unit, a condition of power deviation will occur. This will not only harm the generation unit but also lead to a large-scale damage of equipment at consumer end [15, 16]. To prevent such issues, several grid codes has specified to limit the frequency and provide an acceptable range of operation for the system as depicted in Table 6.2. The system operates with frequency relay [17] which limits the operating range for frequency. When system operates beyond the range the relay disconnects utilities for the DGs. Based on the frequency range, the system is present in time tolerance range for that frequency level. Frequency deviation can cause a wide scale blackout if not mitigated within the prescribed time limit.

Freq (Hz)	England	Spain	France	Germany	Ireland	Denmark	Italy	Sweden	India
53.5 – 52.5				Automatic disconnected					
52 – 51			60 min at 90% power	30 min	60 mins	15 min at 60–100% power		30 min of operation	2s
50.5								Nominal power	
50								Nominal power	
49.5 – 48	Max of 5% active power reduction linearly	3 s	5 h at 90% power / 20 s at 80% power	30 min	60 mins	5 h at 90–100% power / 30 min at 90–100% power / 3 min at 80–100% power		Nominal power / 30 min of operation < 5% power reduction	2s
47.5 – 47	20 s			Automatic disconnected					
46.5 – 46									

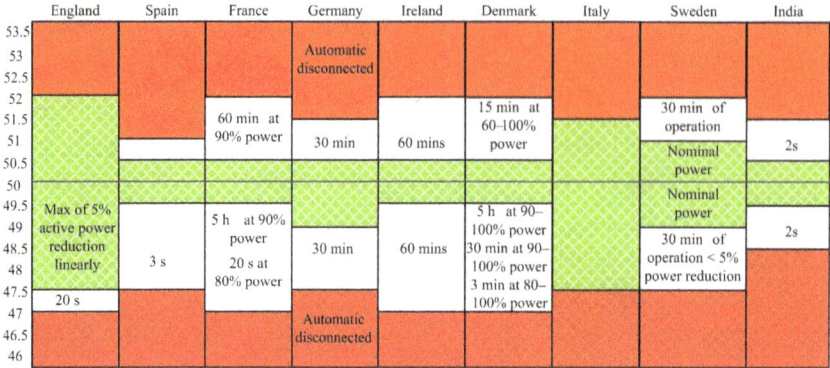

Figure 6.1   Frequency variation as per the grid code of different countries [18]

From Table 6.2 it can be deduced that most of the counties in Asia and Europe operate at 50 Hz nominal frequency whereas, the nominal frequency for North and South American countries is 60 Hz. The permissible range various from ±3% to ±5% in majority of the cases [18]. In most of the European countries there are limit for critical frequency that permits the system to recover within specific time. For instance, in Denmark, the standard operating limit lies from +0.5 Hz to 0.2 Hz. Even if frequency drops below 49.5 Hz and is above 49 Hz, the grid is disconnected only if system does not recover under 5 h of time. Similarly, 30 min of recovery time is provided for frequency which is in the range of 48–49 Hz. A layout to represent the frequency variation range and recovery time is denoted in Figure 6.1.

6.1.3   Power factor, reactive current injection, and reactive power requirement

The reactive power of system directly impacts the voltage at PCC of the GCPVS. Reactive power is supplied to provide stability to the grid voltage during voltage deviation [19]. During grid fault or voltage sag, reactive power is supplied to provide static grid support and reactive current is injected for dynamic grid support [20]. At times, the DGs remain connected with the grid during fault condition for certain period as discussed in previous sections. Few grid code specify that during faulty condition the DGs must inject reactive power for supporting grid voltage recovery. As per German grid code, the reactive current injection needs to be initiated as soon as the grid fault is detected for providing support to grid voltage [21]. A specific amount of reactive current is injected into the grid depending upon the value of voltage sag. In general, ±10% variation in nominal voltage at PCC is acceptable [22, 23]. Hence, when system operates between 0.9 p.u. and 1.1 p.u., the system is in steady state. When the voltage drops below 10% of the nominal value and is still greater than 50% of nominal value [24], reactive current equivalent to 2% of rated current is injected into the grid [25, 26] for every percent drop in voltage. If the voltage drops below 50% of grid rated voltage, then the DGs must inject reactive power equivalent to the rated current. As a result, above explanation can be deduced into following equation:

$$I_q = \begin{cases} 0 & 0.9 \leq V < 1.1 \text{ p.u.,} \\ k\dfrac{V - V_o}{V_n} + 2 & 0.5 \leq V < 0.9 \text{ p.u.,} \\ -I_n + I_{q0} & V < 0.5 \text{ p.u.,} \end{cases} \tag{6.1}$$

where V_n and I_n are nominal voltage and current, respectively. Voltage prior to fault is denoted by V_o and V is voltage during fault condition. The reactive current prior to fault is denoted by I_{q0}.

Recently, many of the GCPVS demands DGs to control the reactive power and utilizing it for stabilizing the grid voltage. On the basic of active power present at any instance the reactive power is controlled. The range of power factor and its relationship with active power and voltage is presented in Figure 6.2.

Based on the grid required of different countries, different grid code specify power factor requirement as presented in Table 6.3.

The requirement of reactive power varies in different countries depending upon the grid codes. In UK for 100% to 50% of active power production, a reactive power of −32% to +32% of rated active power is required. For 50% to 20% of active power production, a reactive power of −12% to +32% of rated active power is required. And for 20% to 0% of active power production, a reactive power of −5% to +5% of rated active power is produced. Whereas in Ireland, for 100% to 12% of active power production, a reactive power of −33% to +33% of rated active power is required. In Germany, the reactive power requirement is based on voltage fluctuation at PCC. When there is a voltage fluctuation from minimum operating voltage to full voltage capacity than the power factor varies from 0.95 inductive to 0.92 capacitive.

6.1.4 Overview of LVRT requirement based on grid codes

In Sweden, for a system capacity less than 100 MW, the DG system shall not be disconnected from the grid for a duration of 250 ms if the system voltage is below 25%. The DG must be capable of supplying 90% of the rated voltage within 250 ms. Similarly, for a system greater than 100 MW, the DG shall not be disconnected from the grid for the duration of 250 ms if the system voltage drops to zero. The DG shall recover 90% of system voltage within 750 ms. According to the Polish LVRT requirements, the system shall be disconnected if the voltage drops below 15% of nominal voltage. The maximal time before beginning of the recovery voltage is 600 ms and the time before which a voltage recovering process should finish is 3 s as represented in Figure 6.3. After this time, the line voltage should reach 80% of the nominal voltage. In UK, the grid regulation for LVRT require that the system should be connected to the grid for a total fault clearance time of 140 ms at zero voltage. Following the fault clearance, the voltage should reach 90% of the nominal voltage within 0.25 s as represented in Figure 6.3. For Danish grid code the DG shall not be disconnected from the grid for the duration of 100 ms if the system voltage is under 20%. The DG shall recover 5% of the system voltage

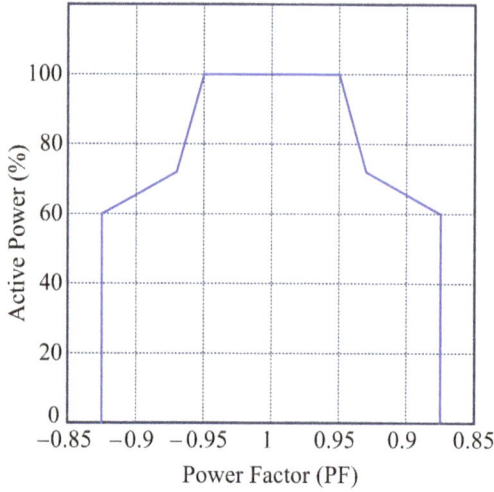

(a)

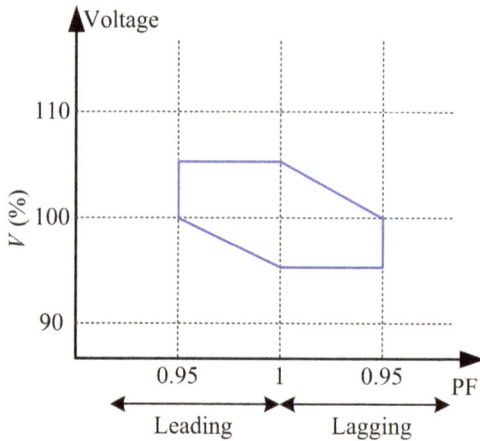

(b)

Figure 6.2 *(a) Range of power factor variation with respect to active power and (b) range of power factor variation with respect to voltage [27]*

within 750 ms. Similarly, for Italian grid code, the DG shall not be disconnected from the grid for the duration of 500 ms if the system voltage is under 20%. The DG shall recover 5% of the system voltage within 750 ms. For Belgium and France, the system should be connected to grid at zero voltage for 200 ms and 150 ms the nominal voltage within 750 ms as per the Belgian LVRT standards, and the grid voltage shall

Table 6.3   Grid code-based power factor requirement

Country	Grid code	Requirement
United States	Federal Energy Regulatory Commission (2005) [5]	0.95 lag to 0.95 lead
United Kingdom	National Grid (2010) The grid code [6]	0.95 lag to 0.95 lead
Ireland	EirGrid [7] EirGrid grid code [7]	0.95 lag to 0.95 lead at full production whereas active power drops to 12% when 100%
China	China's National Standard: GB/T 19963-2011 [8]	0.95 lag to 0.95 lead
Germany	German e-on Grid code [9]	For reactive power injection the power factor must vary from 0.95 p.u. (inductive) to 1 p.u.
India	CERC (Indian Electricity Grid Code) Regulations, 2010 [10]	0.8 lag to 0.95 lead
Denmark	Technical regulation 3.2.1 for power plants up to and including 11 kW [11]	The power factor must range from 0.9 p.u. to 1 p.u., where the active power is 20% greater than the rated power
Australia	Service and Installation Rules of New South Wales [12]	Should not be less than 0.9 lagging

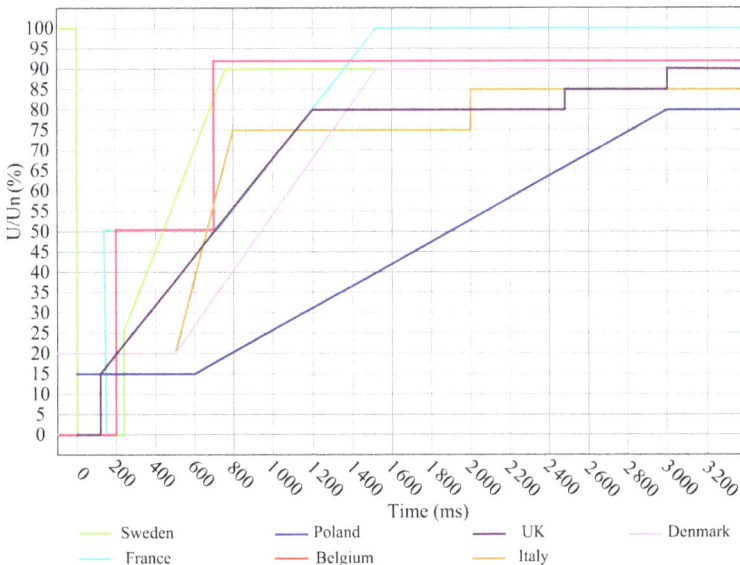

Figure 6.3   LVRT plot based on different grid code [21]

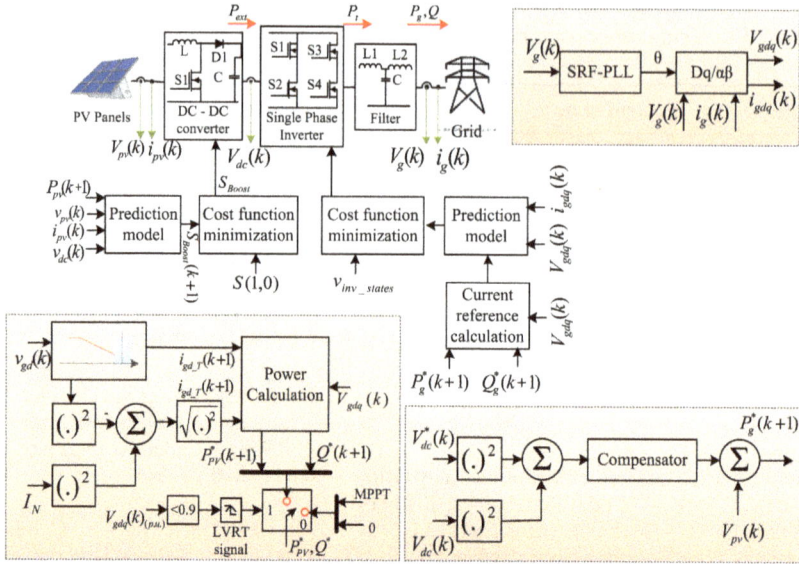

Figure 6.4   *Schematic of the control structure for a single-phase two-stage grid-connected PV system*

restore completely up to 100% of the nominal voltage within 0.15 s as per the French LVRT standards.

6.2   Control strategy used to meet LVRT standards

In a GCPVS, the optimal operation of PV panels is achieved with the help of two typical topologies. These topologies are differentiated by adapting a DC/DC conversion stage between the PV strings and grid tied converter. In the absence of a DC/DC converter, the PV array is directly connected with the grid tied inverter. This approach has disadvantages as the inverter tends to draw a double line frequency current ripple resulting in deviation of the PV voltage away from the voltage at maximum power point (V_{mpp}). Hence, in this chapter, a two-stage topology for grid-tied PV systems is utilized. The two-stage operation of grid tied PV system along with the control layout is illustrated in Figure 6.4.

The two-stage operation provides an additional advantage with the boost converter operating in three different modes. These modes were selected based on the amount of active power to be injected for three voltage regions: (a) the DC/DC converter using MPPT mode operates only on the grids' voltage normal operation region; (b) the reduced power mode (RPM) operates when the voltage sag is between 0.9 p.u. $\leq v_g \leq$ 0.5 p.u., where the PV power is tracked to maintain the inverter power operation; and finally, (c) the short-circuit current mode is activated for severe voltage dips. In the short-circuit current mode, the PV inverter enter to a short-circuit region of the PV panel when the system is required to deliver only reactive power. Considering the operation of the grid-connected power systems with multivariable control algorithms (MCAs) [28, 29], the LVRT operation is achieved as discussed in Sections 6.2.1 and 6.2.4.

6.2.1 Generator-side converter

For a boost converter operating in two stage grid-connected system, the current i_{LB}, when both states of the converter are considered is given by

$$\frac{di_{LB}}{dt} = \frac{1}{L_B}\left(V_{PV} + \left(S_{Boost} - 1\right)V_{dc}\right) \tag{6.2}$$

where V_{dc} represents the voltage at the load R, and S_{Boost} denotes the switching state of the boost converter that can be either 1 or 0, which indicates the states on and off, respectively. The continuous-time model can be converted to discrete-time by approximating the derivative using the forward Euler method [30–32].

$$\frac{dx}{dt} = \frac{x\left(k+1\right) - x\left(k\right)}{T_{sB}} \tag{6.3}$$

where T_{sB} represents the discretization time sample. By substituting (6.3) into (6.2), the discrete-time model of the current i_{LB} is given by

$$\frac{i_{LB}\left(k+1\right) - i_{LB}\left(k\right)}{T_{sB}} = \frac{1}{L_B}\left[V_{PV}\left(k\right) + \left(S_{Boost}\left(k\right) - 1\right)V_{dc}\left(k\right)\right] \tag{6.4}$$

The predicted current in $(k+1)$ can be simplified to

$$i_{LB}\left(k+1\right) = \frac{T_{sB}}{L_B}\left[V_{PV}\left(k\right) + \left(S_{Boost}\left(k\right) - 1\right)V_{dc}\left(k\right) + i_{LB}\left(k\right)\right] \tag{6.5}$$

The presence of C_{PV} in the system helps to maintain the PV voltage constant during operation, and if T_{sB} is small enough, it can be assumed that $V_{PV}\left(k\right) \cong V_{PV}\left(k+1\right)$. Therefore, to obtain P_{PV}, both sides of (6.5) are multiplied by V_{PV}

$$P_{PV}\left(k+1\right) = \frac{T_{sB}}{L_B}V_{PV}\left(k\right)\left[V_{PV}\left(k\right) + \left(S_{Boost}\left(k\right) - 1\right)V_{dc}\left(k\right)\right] + P_{Pv}\left(k\right) \tag{6.6}$$

The variables expressed in (6.5) consider only a one-step prediction, where all the variables can be obtained by measuring the signals indicated in Figure 6.4. Additionally, it does not consider the power required for the charging/discharging of the capacitor. The predicted PV power can be modified to include capacitor dynamics as

$$\begin{aligned} P_{PV}\left(k+1\right) = &\frac{T_{sB}V_{PV}\left(k\right)}{L_B}\left[V_{PV}\left(k\right) + \left(S_{Boost}\left(k\right) - 1\right)V_{dc}\left(k\right)\right] \\ &+ P_{PV}\left(k\right) - C_{PV}V_{PV}\frac{dV_{PV}}{dt} \end{aligned} \tag{6.7}$$

where the derivative can be discretized also using Euler methods. However, the same equation can be rewritten to consider a different prediction horizon h:

$$P_{PV}(k+h) = \frac{T_{sB}}{L_B} V_{PV}(k+h-1)\left[V_{PV}(k+h-1)\right.$$
$$+ (S_{Boost}(k+h-1) - 1) V_{dc}(k+h-1)$$
$$+ P_{PV}(k+h-1) - C_{PV}V_{PV}(k+h-1)\frac{\Delta V_{PV}}{\Delta t_h}$$

(6.8)

In general, the models in (6.5) to (6.8) correlate the P_{PV} output power to the state of the switching signal given by S_{Boost}, and the output voltage of the boost converter. Therefore, one can use (6.8) to predict the PV power, and in turn regulate P_{PV} accurately. Similarly, (6.8) can be solved in terms of S_{Boost} as

$$S_{\_Boost(k+h)=1+} \frac{1}{V_{dc}(k+h-1)} \left\{ \frac{L_B}{T_{sB}V_{PV}(k+h-1)}[P_{PV}(k+h) - P_{PV}(k+h-1)] \right.$$
$$\left. + P_{CPV}] - V_{PV}(k+h-1) \right\}$$

(6.9)

where P_{CPV} represents the capacitor charging/discharging power. In the notation given by (6.9), the predicted P_{PV} has a different meaning. In this case, $P_{PV}(k+h)$ is no longer a predicted signal, but the power reference to track. Therefore, to avoid confusion, (6.9) is rewritten as

$$S_{\_Boost(k+h)=1+} \frac{1}{V_{dc}(k+h-1)} \left\{ \frac{L_B}{T_{sB}V_{PV}(k+h-1)} \left[P^*_{PV}(k+h) - P_{PV}(k+h-1)\right] \right.$$
$$\left. + P_{CPV} - V_{PV}(k+h-1) \right\}$$

(6.10)

Equation (6.10) shows the optimal state of the boost converter that tracks, in this case, the PV power reference.

6.2.1.1 Cost function

As explained in previous sections, the predictive control scheme utilizes an optimization function that executes an algorithm in charge of finding a control action that delivers minimum error. Here, (6.8) and (6.10) provide two possible options to implement MCA; however, they differ in the cost function definition. When using the approach shown in (6.8), the main objective is to regulate the power delivered by the PV. On the other hand, when using (6.9), the cost function objective is changed to find which of the two possible states is closer to the optimal state, which in turn will regulate the PV power. When using the model given by (6.8), the cost function is defined as

$$g_B(k) = \left[P_{PV}(k+h) - P^*_{PV}(k+h)\right]^2$$

(6.11)

where P^*_{PV} represents the PV power reference, in which its h-state can be estimated via an extrapolation technique such as Lagrange or vector angle methods [33, 34]. It is important to mention that for small enough sampling period, no extrapolation is required if the h-state does not exceed 2 [35, 36]. Moreover, in this case, no extrapolation is necessary because the signal reference is constant [37].

On the other hand, in the case of (9), the cost function is given by

$$g_B(k) = \left[s_{Boost}(k + h - 1) - s_{Booststates} \right]^2 \tag{6.12}$$

where $s_{Booststates}$ represents a vector of the states of the boost converter (0 and 1).

6.2.1.2   Control algorithm

Two control approaches can be implemented to track the PV power and they are shown in Figure 6.5. The inductor current, PV voltage, and the voltage output of the boost converter are required for both predictive algorithms (refer to (6.8) and (6.10)). Although both methods utilize a different approach to calculate the optimized control action, they deliver the same result. The main reason to utilize the state-based MCA approach over the conventional method is that the former method utilizes fewer calculations to reach the same result, as can be seen by examining both flowcharts. The cost and minimization functions are calculated the same number of times; however, the predicted step is only calculated once on the state-based technique. Although there is not a significant improvement for the generator-side controller, only one computation was saved and the benefits of utilizing the state-based approach can be seen clearly for power converters that present multiple states like two-level and three-level converters.

6.2.2   Grid-side converter

6.2.2.1   Inverter modeltationary reference

Modeling the inverter shown in Figure 6.4 using phase lock loop (PLL) offers an advantage for the computationally demanding MCA scheme. First, the single-phase variables are transformed from the natural frame to the dq-frame. Second, dq variables are naturally DC quantities during symmetrical conditions, which in turn allows the use of approximations like $x(k) \cong x(k + 1)$, where x represents a state variable. For a single-phase system, the control signals forming the switching states define the value of the inverter terminal voltages as

$$v_i = s V_{dc} \tag{6.13}$$

For a two-level three-phase converter, three control signals labeled S_a, S_b, and S_c are available. These control signals form a total of switching states, which in turn define the value of the inverter terminal voltages as

$$v_{ia} = s_a V_{dc} \tag{6.14}$$

$$v_{ib} = s_b V_{dc} \tag{6.15}$$

$$v_{ic} = s_c V_{dc} \tag{6.16}$$

The grid currents can be expressed in terms of the inverter voltage, the grid voltage, and the inverter filter in the natural frame as

$$\frac{di_g}{dt} = \frac{1}{L_g}(V_i - V_g - R_g i_g) \tag{6.17}$$

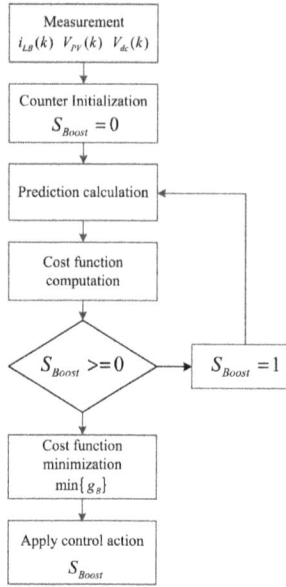

(a)

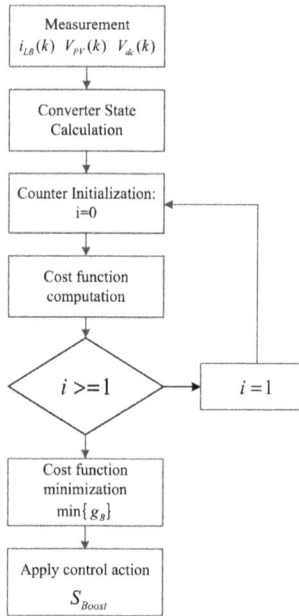

(b)

Figure 6.5   Control algorithm for generator-side stage:(a) power reference-based, (b) converter state-based

where for a three-phase system, $i_g = \begin{bmatrix} i_a & i_b & i_c \end{bmatrix}'$, $V_i = \begin{bmatrix} v_{ia} & v_{ib} & v_{ic} \end{bmatrix}'$, and $V_g = \begin{bmatrix} v_{ga} & v_{gb} & v_{gc} \end{bmatrix}'$.

The grid current dynamics in the natural frame can be converted to the synchronous dq-frame and expressed in state-space form as shown in the following equations:

$$\frac{d}{dt}\begin{bmatrix} i_{gd} \\ i_{gq} \end{bmatrix} = A \begin{bmatrix} i_{gd} \\ i_{gq} \end{bmatrix} + B_i \begin{bmatrix} V_{id} \\ V_{iq} \end{bmatrix} + B_g \begin{bmatrix} V_{gd} \\ V_{gq} \end{bmatrix} \tag{6.18}$$

where

$$A = \begin{bmatrix} -\dfrac{R_g}{L_g} & \omega_g \\ -\omega_g & -\dfrac{R_g}{L_g} \end{bmatrix}; B_i = \begin{bmatrix} \dfrac{1}{L_g} & 0 \\ 0 & \dfrac{1}{L_g} \end{bmatrix}; B_g = \begin{bmatrix} \dfrac{-1}{L_g} & 0 \\ 0 & \dfrac{-1}{L_g} \end{bmatrix} \tag{6.19}$$

where ω_g represents the grid electrical frequency.

The discrete time representation for d- and q-axis currents can be obtained from the continuous-time state-space representation in (6.18) for a one-step prediction, as follows:

$$\begin{bmatrix} i_{gd}(k+1) \\ i_{gq}(k+1) \end{bmatrix} = \Phi \begin{bmatrix} i_{gd}(k) \\ i_{gq}(k) \end{bmatrix} + \Gamma_i \begin{bmatrix} V_{id}(k) \\ V_{iq}(k) \end{bmatrix} + \Gamma_g \begin{bmatrix} V_{gd}(k) \\ V_{gq}(k) \end{bmatrix} \tag{6.20}$$

where

$$\Phi = e^{AT_s}, \quad \Gamma_i = A^{-1}\left(\Phi - I_{2\times2}\right)B_i$$
$$\Gamma_g = A^{-1}\left(\Phi - I_{2\times2}\right)B_g \tag{6.21}$$

where I represents the identity matrix. As it is shown in (6.20), the predicted behaviour of the grid-side currents depends on the present grid variables and the inverter voltages ($V_{id}(k)$ and $V_{iq}(k)$).

Similar to the approach followed in Section 6.1, can be expressed in terms of the control actions, which in this case is the inverter voltage. Therefore, (6.20) can be rewritten as

$$\begin{bmatrix} V_{id}(k) \\ V_{iq}(k) \end{bmatrix} = \Psi \begin{bmatrix} i_{gd}(k+1) \\ i_{gq}(k+1) \end{bmatrix} - \Upsilon_1 \begin{bmatrix} i_{gd}(k) \\ i_{gq}(k) \end{bmatrix} - \Upsilon_2 \begin{bmatrix} V_{gd}(k) \\ V_{gq}(k) \end{bmatrix} \tag{6.22}$$

where

$$\Psi = \Gamma_i^{-1}$$
$$\Upsilon_1 = \Gamma_i^{-1}\Phi \tag{6.23}$$
$$\Upsilon_2 = \Gamma_i^{-1}\Gamma_g$$

where the superscript notation (-1) denotes the matrix inverse. The predicted d- and q-axis grid currents in (6.20) have a different meaning in (6.22). In this notation, the predicted currents represent the d- and q-axis reference currents. Therefore, to avoid confusion, (6.22) is rewritten as

$$\begin{bmatrix} V_{id}(k) \\ V_{iq}(k) \end{bmatrix} = \Psi \begin{bmatrix} i^*_{gd}(k+1) \\ i^*_{gq}(k+1) \end{bmatrix} - \Upsilon_1 \begin{bmatrix} i_{gd}(k) \\ i_{gq}(k) \end{bmatrix} - \Upsilon_2 \begin{bmatrix} V_{gd}(k) \\ V_{gq}(k) \end{bmatrix} \qquad (6.24)$$

where superscript $*$ indicates a reference signal.

Equation (6.24) shows that the inverter control signal at k depends on the current state variables and the reference current to track. Moreover, the matrices Ψ, Υ_1, and Υ_2 can be calculated beforehand to improve the algorithm execution speed.

6.2.2.2 Cost function

Similar to the cost function definition presented above for the generator-side controller, (6.20) and (6.24) allow MCA implementation for inverter control that deliver the same result. When using the model given by (6.20), the cost function is defined as

$$g_g(k) = \left[i_{gd}(k+1) - i^*_{gd}(k+1) \right]^2 + \left[i_{gq}(k+1) - i^*_{gq}(k+1) \right]^2 \qquad (6.25)$$

where the reference and the grid current are compared. On the other hand, if (6.24) is used instead, the cost function is given by the state-based approach as

$$g_g(k) = \left[V_{id}(k) - V_{INV_{States_d}} \right]^2 + \left[V_{iq}(k) - V_{INV_{States_q}} \right]^2 \qquad (6.26)$$

where $V_{INV_{States_d}}$ and $V_{INV_{States_q}}$ represent all possible voltage vectors delivered by the inverter on dq-frame, given by

$$\begin{bmatrix} V_{INV_{States_d}} \\ V_{INV_{States_q}} \end{bmatrix} = \begin{bmatrix} T_{dq} \end{bmatrix} \begin{bmatrix} v_i \end{bmatrix} \qquad (6.27)$$

For a three-phase system, all possible and voltage vectors delivered by the inverter on dq-frame, given by

$$\begin{bmatrix} V_{INV_{States_d}} \\ V_{INV_{States_q}} \end{bmatrix} = \begin{bmatrix} T_{dq} \end{bmatrix} \begin{bmatrix} v_{ix} \end{bmatrix} \qquad (6.28)$$

where $x = a$, b, $c.v_{ix}$ is a $\begin{bmatrix} 3 \times 7 \end{bmatrix}$ matrix that holds all possible voltage vectors that the inverter can provide according to the switching states. Therefore, $V_{INV_{States_x}}$ is a $\begin{bmatrix} 2 \times 7 \end{bmatrix}$ matrix.

6.2.2.3 Current tracking control algorithm and simulation

The current tracking control can be implemented following these steps:

- *Measurement:* The grid voltage, V_g the grid currents, i_g and the DC-link voltage V_{dc} are measured.
- *$\alpha\beta \rightarrow dq$transformation:* Using the estimated grid voltage angle provided by the PLL.
- *Calculation of the converter voltage:* The - and -axs current reference is used in (6.24) for a single computation of the required converter voltages.

- *Calculation of the inverter voltage vectors:* Build the matrix of voltage vectors that is related with the converter switching states using (6.27).
- *Cost function computation:* Because the state-based approach was chosen to calculate the converter voltage, the cost function is calculated using (6.26).
- *Cost function minimization:* Finally, the optimal control action is applied to the converter according to the minimum error of the cost function calculation.

6.2.3   Control structure on symmetrical faults

In the above sections, the details for the control of the generator and grid-side components of the PV system using MCA approach were given. In this section, both controllers are combined in a single control scheme that allows the LVRT for PV systems. Equation (6.24) shows the prediction function used to obtain the converter terminal voltages based on the inverter current references, grid variables, and system parameters. However, the control strategy utilized here is based on the power balance of the DC- and AC-side of the inverter by means of the DC-link voltage control. Therefore, the inverter current references for a single-phase system must be calculated based on the active and reactive power required during the fault event:

$$
\begin{bmatrix} P \\ Q \end{bmatrix} = \begin{bmatrix} V_{gd} & V_{gq} \\ V_{gq} & -V_{gd} \end{bmatrix} \begin{bmatrix} i_{gd} \\ i_{gq} \end{bmatrix}
\tag{6.29}
$$

for a three-phase system are calculated based on the active and reactive power required during the fault event:

$$
\begin{bmatrix} P \\ Q \end{bmatrix} = \frac{3}{2} \begin{bmatrix} V_{gd} & V_{gq} \\ V_{gq} & -V_{gd} \end{bmatrix} \begin{bmatrix} i_{gd} \\ i_{gq} \end{bmatrix}
\tag{6.30}
$$

For a single-phase system, the system described in (6.29) can be utilized to calculate the inverter active current reference for the next sampling time by inverting the matrix:

$$
\begin{bmatrix} i_{gd}^* (k+1) \\ i_{gq}^* (k+1) \end{bmatrix} = \left(\frac{1}{V_{gd}^2(k) + V_{gq}^2(k)} \right) \begin{bmatrix} V_{gd}(k) & V_{gq}(k) \\ V_{gq}(k) & -V_{gd}(k) \end{bmatrix} \begin{bmatrix} P_g^*(k+1) \\ Q_g^*(k+1) \end{bmatrix}
\tag{6.31}
$$

For a single-phase system, the system described in (6.30) can be utilized to calculate the inverter active current reference for the next sampling time by inverting the matrix:

$$
\begin{bmatrix} i_{gd}^* (k+1) \\ i_{gq}^* (k+1) \end{bmatrix} = \frac{2}{3} \left(\frac{1}{V_{gd}^2(k) + V_{gq}^2(k)} \right) \begin{bmatrix} V_{gd}(k) & V_{gq}(k) \\ V_{gq}(k) & -V_{gd}(k) \end{bmatrix} \begin{bmatrix} P_g^*(k+1) \\ Q_g^*(k+1) \end{bmatrix}
\tag{6.32}
$$

where $P_g^*(k+1)$ is given by the DC-link voltage compensator and $Q_g^*(k+1)$ is obtained following grid code requirement and voltage depth.

At the same time, the generator-side controller must operate the PV array at a specific location in the P_{PV} vs V_{PV} curve to avoid grid over current and eventual disconnection from the grid. Therefore, the PV power reference for a single-phase system is calculated based on the future grid current as

$$P_{PV}^*(k+1) = \left[V_{gd}(k) \, i_{gd\_T}(k+1) + V_{gq}(k) \, i_{gq\_T}(k+1) \right] \tag{6.33}$$

The PV power reference for a three-phase system is calculated based on the future grid current as

$$P_{PV}^*(k+1) = \frac{3}{2} \left[V_{gd}(k) \, i_{gd\_T}(k+1) + V_{gq}(k) \, i_{gq\_T}(k+1) \right] \tag{6.34}$$

where i_{gd_T} and i_{gq_T} are the active/reactive currents obtained as per the requirement.

Finally, the control objectives for each stage include the following:

The implementation procedure is illustrated in Figure 6.6 and is as follows: (a) Variables Measurement: Grid voltages and currents, as well as PV voltage, PV current, and DC-link voltage are measured; (b) Synchronous Transformation: the grid voltage and currents are transformed to facilitate the implementation; (c) Fault Detection: V_{gd} is utilized to calculate the presence and depth of a voltage sag; (d) Tracking Reference Calculation: If a voltage sag is detected, the future reference signals $P_{PV}^*(k+1)$ and $P_g^*(k+1)$ are computed warranting the operation limits of the PV system; (e) Prediction Calculation: The predictive models for the generator- and grid-side are utilized to process the next step; (f) Minimization: A cost function is utilized to achieve the minimum error; and, finally, (g) the switching state is applied to both stages of the PV system.

The overall control structure, as well as the two-stage PV system is shown in Figure 6.4. It is important to mention that the active and reactive powers are considered positive when flowing from the DC-side to the grid as it can be observed in the figure. Further, for a three-phase system, the need to control the unbalanced faults is a part of LVRT control.

6.2.4   Control structure and minimization function under unbalanced faults

Severe and abrupt disturbances on the grid voltage have a major impact on the performance of the grid-side converter, especially if the fault is non-symmetrical. When the faults are unbalanced, the injected currents using the approach described in the Section 2.3 are non-sinusoidal, and the active and reactive power includes oscillatory components that are reflected on the DC-link voltage control. These impacts are explained by the presence of a negative sequence component that does not exist during balanced conditions. This new component makes it more difficult and less accurate to retain synchronization with the grid, which in turns deteriorates the output current and DC-link voltage control. According to instantaneous power theory,

(a)

(b)

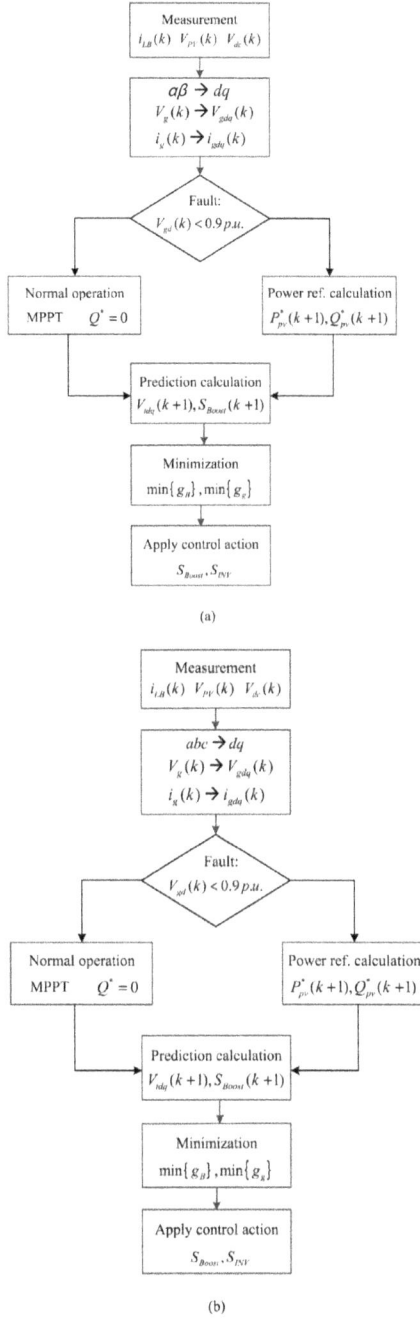

Figure 6.6   Flowchart of the PV system implementation using MCA: (a) control algorithm for single-phase system and (b) control algorithm for three-phase system

and considering only the fundamental frequency, the instantaneous active and reactive power can be defined as

$$p = P_0 + P_{c2}\cos(2\omega t) + P_{s2}\sin(2\omega t)$$
$$q = Q_0 + Q_{c2}\cos(2\omega t) + Q_{s2}\sin(2\omega t)$$

(6.35)

where P_0 and Q_0 are the average values for the active and reactive power, respectively, and P_{c2}, P_{s2}, Q_{c2}, and Q_{s2} represent the second harmonic sine and cosine components of the active and reactive powers that appear when the three-phase system is not symmetrical and balanced. The voltages and currents to calculate the coefficients presented in (6.35) can be expressed in matrix notation using the synchronous reference as

$$
\begin{bmatrix}
P_0 \\
P_{c2} \\
P_{s2} \\
Q_0 \\
Q_{c2} \\
Q_{s2}
\end{bmatrix}
= \frac{3}{2}
\begin{bmatrix}
v_d^+ & v_q^+ & v_d^- & v_q^- \\
v_d^- & v_q^- & v_d^+ & v_q^+ \\
v_q^- & -v_d^- & -v_q^+ & v_d^+ \\
v_q^+ & -v_d^+ & v_q^- & -v_d^- \\
v_q^- & -v_d^- & v_q^+ & v_d^+ \\
-v_d^- & -v_d^- & v_d^+ & v_d^+
\end{bmatrix}
\begin{bmatrix}
i_d^+ \\
i_q^+ \\
i_d^- \\
i_q^-
\end{bmatrix}
$$

(6.36)

where the superscripts $+$ and $-$ represent the positive and negative components of the voltage that can be obtained using PLL. The system of equations in (6.36) and a chosen control strategy can be used to find the positive and negative current references in the predictive model. One possible control strategy would be to eliminate the oscillations on the active power; in this case, the positive and negative current references are given by

$$
\begin{bmatrix}
i_d^{+*} \\
i_q^{+*} \\
i_d^{-*} \\
i_q^{-*}
\end{bmatrix}
= \frac{2}{3}[M_{4\times4}]^{-1}
\begin{bmatrix}
P_0 \\
P_{c2} \\
P_{s2} \\
Q_0
\end{bmatrix}
$$

(6.37)

where the matrix $M_{4\times4}$ includes only the four first rows of the matrix in (6.36). The variables P_0 and Q_0 are obtained based on the DC-link voltage regulation and the sag depth, respectively, Section 6.2.3, whereas the oscillatory components P_{c2} and P_{s2} are equal to zero to eliminate active power oscillations.

The negative current references given by (6.37) are used in the predictive model of the system under its negative sequence components. Therefore, the dynamics of the negative sequence current must be derived to obtain a predictive model similar to the positive sequence calculation shown in (6.24). In this case, the negative components of the injected currents are given by

$$
\frac{d}{dt}
\begin{bmatrix}
i_{gd}^- \\
i_{gq}^-
\end{bmatrix}
= A_{neg}
\begin{bmatrix}
i_{gd}^- \\
i_{gq}^-
\end{bmatrix}
+ B_i
\begin{bmatrix}
V_{id}^- \\
V_{iq}^-
\end{bmatrix}
+ B_g
\begin{bmatrix}
V_{gd}^- \\
V_{gq}^-
\end{bmatrix}
$$

(6.38)

where the superscript – represents the negative component of the electrical parameter, and the matrix A_{neg} is given by

$$A_{neg} = \begin{bmatrix} -\dfrac{R_g}{L_g} & -\omega_g \\ \omega_g & -\dfrac{R_g}{L_g} \end{bmatrix} \tag{6.39}$$

In general, (6.38) follows the same structure of the model presented in (6.18).
Therefore, the analysis done for the positive sequence that goes from (6.20)
to (6.24) to obtain the predictive model must be done as well for the negative
component.

Finally, the cost function definition for the grid-side formulated in (6.26) has to
be modified to include the negative sequence injection objective. Therefore, the cost
function using the state-based approach is formulated as

$$\begin{aligned} g_g(k) &= \left[V_{id}(k) - V_{INV_{States_d}} \right]^2 + \left[V_{iq}(k) - V_{INV_{States_q}} \right]^2 \\ &+ \lambda \left\{ \left[V_{id}^-(k) - V_{INV_{States_d}}^- \right]^2 + \left[V_{iq}^-(k) - V_{INV_{States_q}}^- \right]^2 \right\} \end{aligned} \tag{6.40}$$

where $V_{INV_{States_d}}^-$ and $V_{INV_{States_q}}^-$ represent all possible d and q voltage vectors delivered by
the inverter in the dq-frame rotating in the opposite direction of the positive sequence,
and are given by

$$\begin{bmatrix} V_{INV_{States_d}}^- \\ V_{INV_{States_q}}^- \end{bmatrix} = \left[T_{dq}^{-1} \right] \left[v_{ix} \right] \tag{6.41}$$

where $x = a$, b, $c.v_{ix}$ is a $\left[3 \times 7 \right]$ matrix that holds all possible voltage vectors that
the inverter can provide according to the switching states, and λ is transformation.
On the other hand, represents the weighting factor for the negative current injection.

Finally, the implementation with negative compensation follows the same structure presented in the flowchart of Figure 6.6 with two additional aspects: (a) the
calculation of the predictive negative current component and (b) the minimization
function expressed in (6.40).

6.2.5 Experimental analysis

To assess the robustness of the developed fault ride-through controller, a line to
ground fault is injected in 10 kW single-phase GCPVS. The line fault creates a voltage sag impact on the normal operation of the power system network. The voltage
and current at the PCC during the action of sag fault for a single-phase system are
shown in Figure 6.7. From the figure it can be identified that at $t = 0.15$ s, a voltage

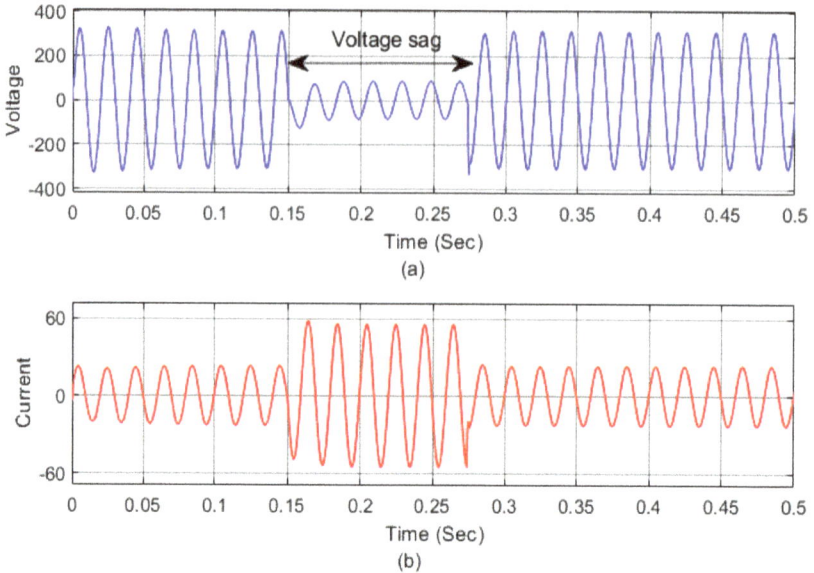

Figure 6.7   *Voltage sag effect on single-phase GCPVS: (a) voltage and (b) current at PCC*

sag occurs disrupting the normal operation of the system. At this instant, the corresponding current at PCC increases up to the threshold limit. During this condition, the inverters control is designed to operate with constant peak current control through the duration of the event to ensure safe operation and avoid over current loading.

The power output of the system during the inverter operation under the voltage sag profile is shown in Figure 6.8. At $t = 0.15$ s, the active and reactive power,

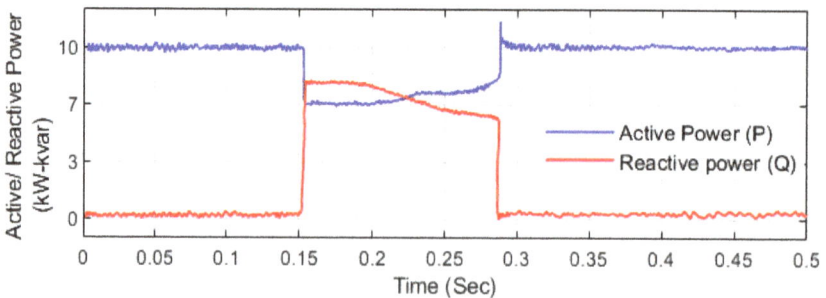

Figure 6.8   *Active and reactive power during voltage sag effect on single-phase GCPVS*

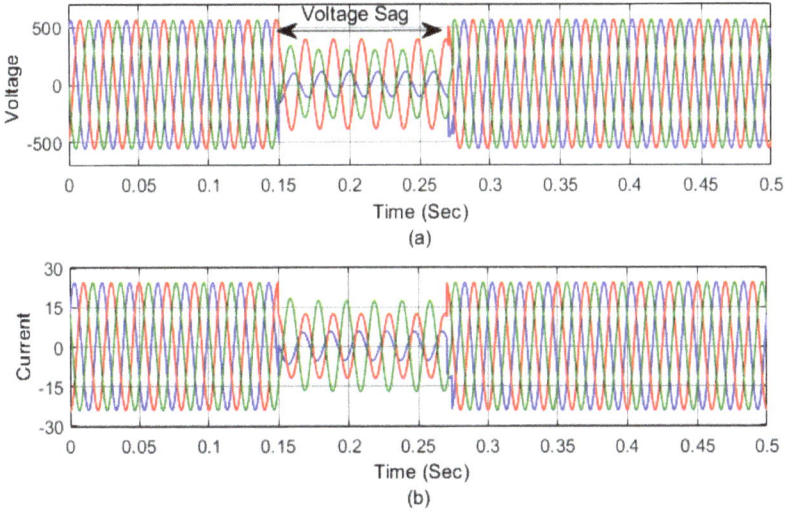

Figure 6.9   Voltage sag effect on single-phase GCPVS: (a) voltage and (b) current at PCC

experience high-frequency oscillations because of negative sequence components, and the PV system is exchanging power with an unbalanced grid system. The average value of PPCC drops from 10 kW to 6.9 kW and a reactive power of 8 KVar is injected with the help of proposed controller. This stabilizes the system voltage and restores the normal operation of the system before 1.35 s.

Similarly, the robustness of the developed fault ride-through controller is tested for a 10 kW three-phase grid-connected system by injecting a line fault in the system. The line fault creates a voltage sag impact on the normal operation of the power system network. The voltage and current at the point of coupling during the action of sag fault for a three-phase system are shown in Figure 6.9. From the figure it can be identified that at $t = 0.15$ s, a voltage sag occurs disrupting the normal operation of the system. At this instant, the corresponding current at PCC increases up to the threshold limit. During this condition, the inverters control is designed to operate with constant peak current control through the duration of the event to ensure safe operation and avoid over current loading.

The power output of the system during the inverter operation under the voltage sag profile is shown in Figure 6.10. At $t = 0.15$ s, the active and reactive power, experience high-frequency oscillations because of negative sequence components, and the PV system is exchanging power with an unbalanced grid system. The average value of active power drops from 10 kW to 5 kW and a reactive power of 6 KVar is injected with the help of proposed controller. This stabilizes the system voltage and restores the normal operation of the system before 1.4 s.

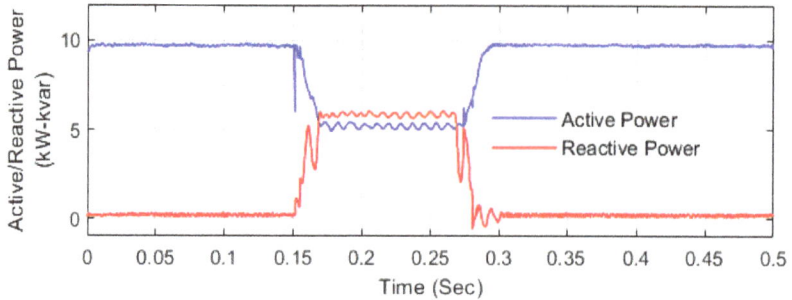

Figure 6.10   Active and reactive power during voltage sag effect on single-phase GCPVS

6.3   Anti-islanding requirements for control

The anti-islanding detection of GCPVS is necessary during the power system failure. If the DGs are unable to recover the grid from fault condition by ride-through, an anti-islanding protection disconnects the DG from the grid and provides a necessary safety to both the utilities as well as the DGs and avoid the complete blackout for grid [38]. The main aim of anti-islanding protection is to assure power system health and make sure that no unintentional islanding has occurred in the local area for safety reason [39]. During unintentional islanding [40], the DGs tend to energies the local area at the PCC. It leads to the islanding of DGs area at local level and the grid is unaware of the islanding that has taken place. As a result, it can be hazardous for the safety personal who is working on line maintenance as he is unaware about the operation at the DGs end and he faces a threat of being electrocuted [41]. Even inadequate grounding during unintentional islanding may lead to large transient over voltage due to rapid loss in load [42]. The change of power quality and transient torque due to out of phase synchronization of machines are some other issue that can be of concern because of unintentional islanding [42]. Hence, communication between DG and the utilities is necessary for avoiding any severe issues [42]. As a result, many countries have come up with anti-islanding standards in their grid codes which specifies when the grid requires to be disconnected and based on different parameters at PCC. The severity of fault helps in setting up the priority and decide the instances at which the DGs disconnect from the grid. A few of the grid standard for anti-islanding are Section 6.3.3.

6.3.1   IEEE Standard for Interconnection and Interoperability of resources with interfaces (IEEE 1547)

The standard was documented by institute of electrical and electronics engineering (IEEE) with an aim to provide a guideline for distributed energy resource (DER) interconnection into the grid. It helps the energy industry to perform business with the different stack holders and influence the future of power system industry. The

IEEE 1547 [43] provides a list of regulations at local stand and federal levels to provide transparency and fairness in implementation DER interconnection with the grid. The major areas of concern during implementation of IEEE 1547 standard are:

IEEE 1547 (2003) [43] was the first edition of the standard concerning over DER interconnection. DER comprises DG as well as energy storage system. The standard focus on technical specification regarding testing and interconnection and does not focus on the technique of DER and stand as technologically neutral. The standard provides significance to operation, testing, performance, and maintenance of interconnection with DER. The conditions such as power quality, islanding state, and response time for abnormal condition are specified.

IEEE 1547 (2003) provided a criteria for interconnection but it lacked in providing any application-based guidelines. Further issue regarding self-protection of DER along with designing and maintenance of DGs and utility grid is not presented in the standards. Hence, the IEEE 1547.1 (2005) was designed to provide the specification for testing and evaluation. IEEE 1547.1 specify the type of test, product to be tested, and method of evaluating the test that is to be performed to establish the interconnection between DER and utility grid. IEEE 1547.2 (2008) [44] provides a background on 2003 standard and includes its logic to provide justification to technical description, application guidelines, schematics and interconnection examples for improving 1547. With rise in smart grid technology IEEE 1547.3 (2007) was designed, which presented a guidance for monitoring and information exchange and controlling via communication network. IEEE 1547 (2003) did not focus on intentional islanding and microgrid-based requirement and with the rise in renewable energy it was vastly required hence in IEEE 1547.4 (2011) the standards was updated. It aims on disconnecting the grid from the DGs in case of grid loss is detected and DG operations in islanding state. It provide good practices which need to be incorporated while designing, operating, and integrating DER with the utility grid. The standards provided instruction for DG to operate separately in islanded form and reconnecting grid once the grid is stable. The standards was revised in 2013 IEEE 1547.7 which address the impact of DG on the grids.

In 2014, the IEEE 1547 (2014) [45] standard was amended, and second issue was published. There has been a lot of development in the file of interconnection of DER, i.e., smart grid advancement, since the first issue was published in 2003. During the amendment, operation of DGs and coordination between DER was considered. By changing the active and reactive power, DER were now allowed to regulate the voltage level. The manufacturer must specify the characteristics of the equipment which will inject active and reactive power for regulating the voltage. Equipment can generally respond to variation of grid voltage which can be identified either by communication setting or by time scheduling. The amendment mostly aimed on flexible operation of the DG to avoid fault ride-through and attain prescribed voltage and frequency level for interconnection DERs. Based on the amendment IEEE P1547.1a (testing procedure) was also amended. The amended standard covered testing for voltage regulation and frequency ride-through. The equipment category for DER voltage regulation was created and functionality of the equipment under testing (EUT) are as follows [46]:

In 2018, the IEEE 1547 [47] was again amended to provide a uniform standard for interoperation and interconnection of DER with electrical power system. The requirement which are relevant for interconnection and interoperability such as performance, testing, operation, maintenance, safety, and security are considered. The revised 1547 (2018) standard addresses:

6.3.2   Utility-interconnected PV inverters—test procedure of islanding prevention measures (IEC 62116)

The standard was documented by international electrotechnical commission (IEC) with an aim to provide a guideline for interconnection of PV inverter with the utility. The first issue was published in 2008 and was latter replaced by updated issue in 2014 [48]. The updated considered active power of the system for most of the clause, whereas 2008 version considered real power. According to the standard, islanding is considered when a portion of utility consisting of both load and generation unit is isolated from the power grid. Islanding is created by controlling the utility and isolating a large section of grid, such action is known as intentional islanding. On the contrary unintentional islanding is created when grid containing load and only consumer owned generation is isolated from the grid by utility control. In regular application the consumer owned generation senses the absence of utility grid and terminate energy supply to the grid. However, when there is a presence of a well-balanced load and DGs, prior isolation and a very small amount of power is being used by the utility. Then it is difficult to detect the isolation of utility taking place. It can be very harmful as utility may not be controlling the generation and it may operate in some other frequency and voltage level causing damage to equipment present at the load terminal. Even the reconnection can lead to damage as the DGs may be out of synchronization from the grid. The energize line at the islanded end also causes threat to the line worker who is assuming that the line is dead from the grid-side.

Based on these issues, PV industry has developed various islanding detection and preventive measure over the years. The test procedure were created to check the efficiency of this detection mechanism as demonstrated in this document. The standard provided a compromise test procedure for evaluation of the islanding detection process along with overview of preventive measure to be taken for power conditioning of interconnection. The standards is specific to solar inverter but with some modification can be utilized for inverters using different power generation sources. The inverters and the other devices that meet the requirement of the document can be considered for the non-islanding operation.

6.3.3   IEEE recommended practice for utility interface of PV systems (IEEE 929)

The standard was documented by IEEE with an aim to provide guidelines regarding the equipment and tasks essential for ensuring smooth operation of PV system which is interconnected with grid in parallel [49]. The factors such as power quality, equipment safety, and personal safety are considered. Recommendation regarding island operation of PV system is also presented.

The recommendation in this documents are generally applied to the PV system which are interconnected with the grid in parallel. The recommendation is for static state inverter that converts DC to AC and is not applicable for rotating inverter. The recommended practices are specifically for small system with 10 kW or less rating which can be generally applicable for individual residences. The standardization of small-scale system tends to reduce the burden over both the utility as well as PV system installer. Medium and large-scale installation may combine various standards and customize as per requirement.

6.3.4 Requirement overview for anti-islanding operation

Islanding condition takes place when the DGs are disconnected from the grid and the generated power is used to feed the local load. It is mostly dominant on the low-voltage side of the network as a result it is required to disconnect the system if the frequency is not in the range of 49–51 Hz [50]. As discussed earlier, the grid codes need to be more stringent and need to be updated regularly as the amount of PV installation is increasing and will keep on expending in the future. When the large PV plant is integrated with the LV network, a large amount of active power is generated which leads to rise in voltage at the feeder and can exceed the specified limit. Prior there was no contribution of PV inverter in grid stability. But German standard VDE0126.1.1 (2015) [51] presented the following scenario when the inverter must disconnect from the grid:

Introduction of smart inverters have made the control process more challenging. Now the inverter are required to contribute to grid stability and support the grid during abnormal operating condition.

It is necessary that all the properties regarding the power quality of the DGs are maintained as per the grid code for interconnection of PV system.

6.4 Control strategy used to meet anti-islanding standards

Islanding is a protection scheme which disconnects the electrical system from the grid and islanded DGs generated power is consumed by the local load [52] Section 6.3. One of the major constant between all the standards is that, in case of any abnormality at the grid end, the islanding detection algorithm must be able to detect it and disconnect the DGs from the grid under 2 s [53]. A few of the challenges that are presented while islanding DGs are as follows [54]:

Hence based on the standards and by keeping challenges in check, many different islanding detection methods have been proposed by the researchers over the years. The islanding detection technique can be classified into three categories: communication-based, local detection method, and intelligent islanding detection [55]. The local detection technique can be further classified into active passive and hybrid detection technique as depicted in Figure 6.11.

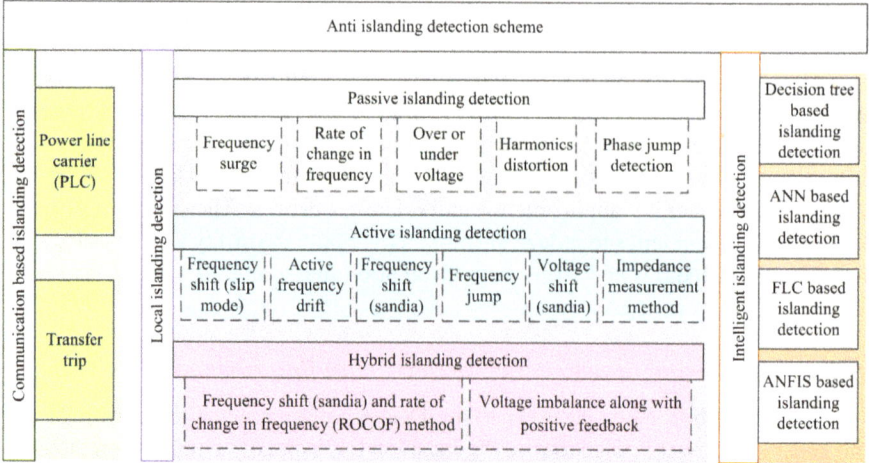

Figure 6.11　Islanding detection schemes

6.4.1　Communication-based islanding detection scheme

It is one of the most reliable islanding detection scheme in which there is a direct communication between DGs and the utilities. The method is reliable but there is an issue regarding complexity and cost of implementation [56]. A few of the commonly implemented communication-based islanding detection techniques are sections.

6.4.1.1　Power line carrier (PLC) communication

The power line is used to carry the communication signal with information regarding DGs state of operation (islanded or non-islanded). A signal generation is present at the substation of 25 kV or higher rating which is coupled with the network and keeps on broadcasting the DGs states information continuously as depicted in Figure 6.12. Because of the low-pass nature of the filter present in power system, it is required that the signal transmit either at or below the fundamental frequency and does not present any interference with other carrier signals [57]. During the normal operation, the signal received by the DGs and ensure the interconnection between

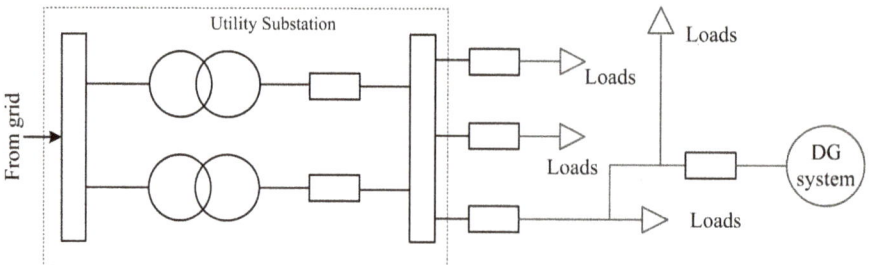

Figure 6.12　PLC islanding detection schemes

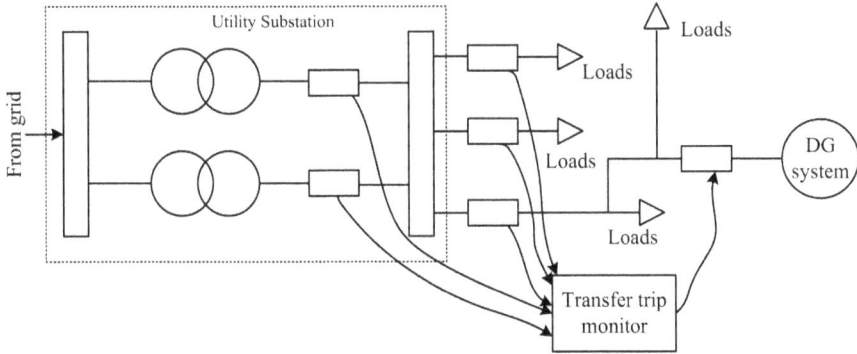

Figure 6.13   Transfer trip islanding detection schemes

utilities and DGs. Whereas in case of fault the signal is cut-off by the substation breaker and DGs stats to operate in islanding mode. The method is highly reliable and is simple in terms of control. For radial system only one transmission generator is required to communicate with multiple DGs continuously. The only time communication is disrupted is when there is a fault in line, or the line has been disconnected by an open breaker. In [57], a brief investigation on PLC-based islanding detection technique is presented and the data is verified on the field.

Even though the technique is vastly reliable and simple to implement, it does present few difficulties during practical implementation. At the substation voltage needs to be transformed from high potential to lower potential with the help of transformer. The transformer may present a cost barrier which may cause the installation of DGs undesirable for local network. In case of non-radial signal, multiple signal generator are required to communicate, which will lead to increment in cost by three folds. Another concern is regarding the transmission signal frequency which need to be near the fundamental frequency value. The energy required to attain such condition is very high and the value of SNR also increase with frequency in communication signal. In previous research it was reported that vibration in motor may lead to voltage fluctuations and in this method the trip signal may be generated [58].

6.4.1.2   Transfer trip

In transfer detection scheme, all the circuit breakers are linked and monitored by a centralized DG control as depicted in Figure 6.13. If a circuit breaker is tripped at any certain instance, the substation determines the islanded area and send a signal to the DG informing about islanded state and mode of operation to be performed further [59]. The transfer trip has a distinct advantage over PLC. With radial connection consisting few DG sources and a limited number of circuit breakers, the state of the system can be monitored by DG at each point [60]. It is also the limitations as for large system the complexity of control and cost of implementation are two major concerns. The size of the system makes the control complex and transfer trip obsolete.

6.4.2   Local islanding detection scheme

In case of local detection, the power producers are independent for detecting abnormality and disconnecting them from the system without any interface from the local utilities. The method is cost effective in comparison to the communication-based islanding detection. Local islanding detection can be further classified into active, passive, and hybrid islanding detection technique. During passive islanding detection scheme [61], the terminal voltage, current, and frequency of the DGs are monitored for any irregularities. Because of sensitivity limitation of the technique active islanding detection techniques was proposed. As per the literature [62] in active islanding detection method a signal is injected in the network by the DG and is constantly monitored for any reaction which is compared with preset threshold value. However, the introduction of external signal may lead to power quality degradation and system can become unstable if a significant amount of signal is injected [63]. To overcome the drawback of both the methods, hybrid islanding detection technique [64] was introduced in which the external signal is injected when abnormality is detected in the system as a result the non-detection zone (NDZ) is small. Nevertheless, the islanding detection time is prolonged as both the detection methods are implemented.

6.4.2.1   Passive islanding detection technique

Passive islanding detection is one of the most widely used islanding detection scheme in which voltage, current, and the phase variation at the PCC is measured and islanding takes place in case of any abnormality [61]. It is one of the most cost-effective method as it utilizes the relay which are already in place for protective measures. The sensor threshold is set to differentiate between the normal mode of operation and operation in islanded state. One of the biggest drawback being carried out in the field of passive islanding detection. The advantages and limitations of passive islanding detection techniques are presented in Table 6.4. The technique can be further classified as:

6.4.2.1.1   Frequency surge/over or under frequency

The frequency-based islanding detection scheme is mostly implemented for induction and synchronous machines. If power mismatch between generated and consumed power occurs, there is a possibility of frequency change for the system [66]. As discussed in Section 6.3 that as per the standards the frequency need to be constant and a variation of 10 kW is permissible. Hence, if the frequency is found beyond the threshold limit then the system is operating in islanding condition and the DG relay open the breaker to avoid any damage. As per the literature [67], the change in power or power imbalance (P_l) is the difference between power generated by DG (P_{dg}) and power consumed by the load ($8KVAR$). By applying the value of P in swing equation, the value of frequency decline can be determined when the generation capacity is lower than the load requirement.

Table 6.4   Comparative analysis of passive islanding detection scheme [65]

Detection method	Advantage	Disadvantage
Frequency surge/over or under frequency	Low cost of implementation Same method can be used for protection loads and equipment from damage	NDZ is large The time of reaction for various protection equation is different
ROCOF	Faster detection speed compare to under and over frequency High sensitivity High efficiency even for small frequency error	The threshold selection is difficult Error may occur as the cause of frequency change is unknown
Over or under voltage	Low cost of implementation Same method can be used for protection loads and equipment from damage	NDZ is large Reaction time varies for different protective equipment
Harmonics distortion	Easy to implement High speed of detection	The threshold selection is difficult NDZ is large Multiple DGs may degrade the performance
Phase jump detection	Easy to implement High speed of detection Multiple DGs have minimal impact on efficiency	The threshold selection is difficult NDZ is large and can cause failure for local load if phase error is not sufficient

$$P = \sum P_{dg} - \sum P_l \tag{6.42}$$

$$\Delta f = \frac{\Delta P}{D}\left(1 - e^{-Dt/2H}\right) \tag{6.43}$$

where the frequency change is indicated by Δf, ration of percentage load change to percentage frequency change (load damping factor) is denoted by D, and system inertia constant is denoted by H.

6.4.2.1.2   Rate of change in frequency

It is another frequency-based method for islanding detection. Instead of Δf, rate of change in frequency (ROCOF) $\left(\frac{df}{dt}\right)$ is used. The DG is set to trip if the value of $\frac{df}{dt}$ is beyond a certain threshold value. In majority of the cases the threshold is set to be 0.1 Hz/s to 1.2 Hz/s [54]. Consideration regarding the start-up frequency variation and fault-based frequency variation need to be made clear to avoid any unwanted trip taking place. The relation between power and voltage (magnitude, phase, and frequency) needs to be clear as per the following equations:

$$P + jQ = \frac{V_{a1}V_{b1}\angle\theta}{Z_1} \tag{6.44}$$

$$P = \frac{V_{a1}V_{b1}}{Z_1}.\sin(\theta) \tag{6.45}$$

$$Q = \frac{V_{a1}}{Z_1} \cdot \left(V_{b1} \cdot \cos\left(\theta\right) - V_{a1}\right) \tag{6.46}$$

where active and reactive power are represented by P and Q, respectively. The voltage at two points are represented by V_{a1} and V_{b1} and impedance of the section is denoted by Z_1. The change in value of P results into frequency and phase variation whereas voltage is varied by change the value of Q.

6.4.2.1.3   Over or under voltage

Passive islanding detection can also be performed by monitoring the voltage of the system. The reactive power mismatch will result in voltage variation. The excess of reactive power will cause over voltage condition whereas the lack of reactive power will lead to under voltage condition. If the voltage falls out of the preset threshold value, the relay will open the breaker and disconnect DG from the utility, initializing islanding state of operation. Most of the studies indicate that the voltage-based islanding detection method has higher performance in comparison to that of frequency-based islanding detection technique [68].

6.4.2.1.4   Harmonics distortion

This methods of islanding detection technique is majorly implemented in system comprising of an inverter, where harmonics are likely to be present. In this method the total harmonic distortion (THD) of the system is measured and the threshold is set. The relay open the breaker if the THD exceed the desired value [69, 70]. Various factor such as motor drive, switching power supply, and nonlinear component are responsible for variation in harmonics. The level of harmonics fluctuates with variation in load. As per the standard discussed in Section 6.3, the permissible limit for THD variation is 5% for voltage waveform and 10% for current waveform [71] under full load condition.

6.4.2.1.5   Phase jump detection

In phase jump-based detection method, the change in phase angle of voltage waveform to that of reference waveform is monitored. In case of fault, an instantaneous phase shift in DG is detected to accommodate the change in power requirement [72]. Threshold based on maximum permissible jump limit is set and if the phase jump is beyond the limit then the relay is triggered. As per the standards, if a power mismatch of more than 33% is present then the relay will disconnect the DG after 300 ms of detection time [73].

6.4.2.2   Active islanding detection technique

For active islanding detection method, a small perturbation is injected into the grid. The generated perturbation will cause an insignificant disturbance when the DGs are connected to the utilities, whereas the disturbance becomes noticeable when DGs are disconnected [62]. The active islanding detection technique has a smaller NDZ which is its advantage over passive techniques. The instability of grid due to large

Table 6.5   Comparative analysis of active islanding detection scheme [65]

Detection method	Advantage	Disadvantage
Frequency shift (slip mode)	NDZ is small Highly efficient even for multiple DGs Present highly efficient detection method, with improved power quality and improved transient stability of the system	Multiple degrees of transient stability and power quality Level of penetration is high, which leads to reducing the stability
Active frequency drift	NDZ is small Easy to implement with the help of microcontroller No NDZ present for resistive load	Degradation in power quality Chopping factor controls the NDZ For multiple DGs the efficiency of the detection method falls
Frequency shift (Sandia)	For single inverter NDZ is almost equal to zero Highly efficiency on being coupled with Sandia voltage shift scheme Present highly efficient detection method, with improved power quality and improved transient stability of the system	Vulnerable to harmonics Degradation in stability and power quality
Frequency jump	Efficient for sophisticated frequency scheme NDZ is negligible for single inverter	Not effective for multi-DGs as frequency synchronization is a issue
Voltage shift (Sandia)	Easy to implement Efficiency of detection improves on implementation along with Sandia frequency shift	Dependent on inverter output as a result easily effected by implemented MPPT algorithm
Impedance measurement method	NDZ is small for a single inverter	If DGs are not operating in synchronization the detection efficiency falls Selection of threshold value is difficult as value of grid impedance is required

number of inverters is one of its prime disadvantage [62]. This technique utilizes advanced controlling method along with load, voltage and current variation which is followed by measuring the detection time to monitor the islanding and non-islanding condition. The advantages and limitations of active islanding detection techniques are presented in Table 6.5. The active islanding techniques can be further classified as follows:

6.4.2.2.1   Frequency shift (slip mode)

In slip mode-based islanding detection technique, a positive feedback control is used which tends to destabilize when islanding condition take place. The method can be explained by the equation

$$i_{inv} = I_{inv} \sin\left(\omega t + \varphi\right) \tag{6.47}$$

For a current source inverter, the phase of positive feedback is represented by φ, Hence, the value of φ needs to be varied for sliding the frequency and causing a short-term frequency change. For implementation of slip mode, modification are made in the PLL filter which lead to out of phase of fundamental value. During normal operating condition PLL track the phase and frequency changes in the network, but during faulty condition the slip mode keeps the inverter in phase. In case of frequency shift during islanding condition the PLL will observe a negative error and will try to shift the frequency away from the fundamental [66]. Because of the positive feedback the phase shift will occur in the reverse direction causing the frequency to go beyond the breaker threshold and causing the relay to disconnect the breaker.

Slid mode is easy to implement as only the value of inverter filter is to be varied. As per the literature [74] slip mode has one of the smallest NDZ and is effective with multiple inverters. However, this method weakens in case of frequency response for RLC load is greater than slip mode-based system.

6.4.2.2.2   Active frequency drift

It is a computer-based islanding detection technique in which frequency of the output is distorted to create a continuous drift of the frequency from the fundamental. By altering the frequency and by incrementing the frequency for each half cycle followed by dead time, the system tends to wait for the fundamental to match the biased frequency [74]. Current for each cycle can be deducted as per the equation

$$i = \sqrt{2I} \sin\left[2\pi\left(f_V + \delta f\right)\right] \tag{6.48}$$

where the output current is deduced by i, maximum current is deduced by I, voltage frequency is represented by f_V and change is frequency is denoted by δf.

6.4.2.2.3   Frequency shift (Sandia)

Sandia-based frequency shift techniques is an advancement of positive feedback-based frequency biased technique. The error from the dead zone is taken by the positive feedback as a frequency error as depicted in Figure 6.14.

When the DG systems are connected to the grid, even a small variation in frequency will lead inverter to go beyond the range of line frequency, but the grid may keep the system stable. From the equation the feedback signal is represented.

$$\theta_f = \frac{\pi}{2}\left(cf_0 + K\left(f - f_0\right)\right) \tag{6.49}$$

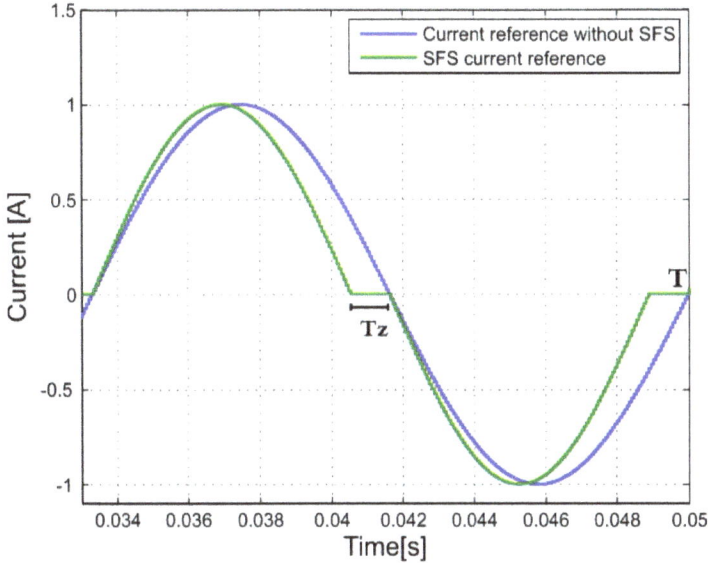

Figure 6.14   Disturbance introduced by frequency shift islanding detection [76]

where the terminal frequency is represented by f and I represents the base frequency (50 Hz), positive feedback gain is depicted by K, and cf_0 represents initial chopping fraction.

When the fault occurs, the frequency error increases along with the dead zone. The method has a significant merit over the frequency bias method [75]. Only the transient response efficiency is reduced when high-density source is present in the system. The method is implemented when a current control inverter is used, and it remains unsuitable for small DG integration [76].

6.4.2.2.4   Frequency jump

Frequency jump method is closely related to frequency biased method as dead zones are added but not for every cycle. Dead zone is added after every second cycle as per the predefined algorithm. On interconnecting the utility with the DGs the inverter current is modified whereas, grid linking voltage is observed. During the islanded condition, the current and voltage varies as per the inverter [74]. As a result, the islanding state can be detected by frequency modification. The efficiency of this method reduces when more than one inverter is interconnected.

6.4.2.2.5   Voltage shift (Sandia)

In Sandia frequency shift method, voltage shift also uses positive feedback for islanding detection. The inverter power is decreased along with voltage during the application of this detection method. During the normal operation, there is no change in output terminal voltage as the grid provide system stability. Whereas, in case of faulty condition there is a drop in voltage along with drop in power [74]. The positive feedback controller further drop the power to a point when the threshold value of relay is exceeded, and the voltage protection relay disconnects the DG from the grid.

6.4.2.2.6 Impedance measurement method
The method focus on fluctuation in voltage and its response on the current. During normal operation when the grid is interconnected with the inverter the change in output current will impact the voltage of the inverter as represented in the equation

$$V = \frac{P_{dg}}{2} \sqrt{\frac{R}{P_{dg}}} \tag{6.50}$$

where P_{dg} represents the active power for the DG, R is the voltage at PCC, and R denotes the resistive load. As the value of R and P_{dg} is constant, the variation in voltage is directly proportional to change in active power. The method has a small NDZ and even if the PV inverter output and load are balanced during the islanding condition, the inverter output tend to vary with the load causing a trip [77]. The presence of multiple inverters reduces the efficiency of the technique as multiple inverters may cause flickering that may lead to false trip signal generation.

6.4.2.3 Hybrid islanding detection scheme
Hybrid islanding detection method comprises of both active and passive islanding detection method. The perturbation for active islanding detection is added to the system when islanding is detected by passive islanding detection algorithm [64]. Multiple passive detection techniques can be used with any one active detection techniques simultaneously. A few of the most used hybrid algorithms are as follows:

6.4.2.3.1 Frequency shift (Sandia) and ROCOF method
In this method of islanding detection Sandia frequency slip is an active islanding detection technique whereas ROCOF is the passive islanding detection technique. Whenever islanding is detected by ROCOF algorithm as explained in section 6.4.2.1 it activates the Sandia frequency shift algorithm to check if the islanding detection by ROCOF is appropriate [63]. The trip signal at multiple point of the DG-grid interconnection is activated by frequency shift algorithm. This method prevents false trip from taking place and solve the issue of large NDZ present in passive islanding detection algorithm.

6.4.2.3.2 Voltage imbalance along with positive feedback
Like previous method, the passive islanding method of voltage imbalance is coupled with active islanding detection technique of positive feedback. The voltage at the PCC of the system is monitored constantly and in case of any abnormality is

observed the frequency set point of the DG is varied. The trip signal that control the relay is activated by the positive feedback [78]. The technique provide a large NDZ but because of operation of two islanding detection technique simultaneously, the operation time for islanding detection is increased.

6.4.3 Intelligent islanding detection scheme

Intelligent islanding-based technique refers to set of algorithms which are trained to imitate response like human intelligence. The techniques tends to learn from the past database and implement action based on trained dataset. A few of the most implemented techniques are artificial neural network (ANN), adaptive neuro fuzzy-based inference system (ANFIS), fuzzy logic control (FLC), and decision tree. The algorithms are equipped to solve multi-objective non-linear problems efficiently and more accurately when compared to conventional algorithms [55]. Some of the implemented techniques are discussed in Sections 6.4.3.1–6.4.3.4.

6.4.3.1 ANN-based islanding detection

ANN-based islanding detection is one of the most used machine-learning-based algorithm for solving various engineering problems. In ANN, network between neurons is represented in the form of biological interconnection between the cells. For power system-based issues, multilayer feed forward network analysis is adopted in general. Various researchers have implemented a neural network (NN)-based islanding detection techniques for multi inverters and hybrid DGs [79]. As per the literature [80], mostly the voltage at PCC is usually monitored and there is no impact on the power quality as in case of passive islanding detection technique. The feature of voltage signal obtained at PCC is extracted by discrete wavelet transform (DWT) [81, 82]. A few of the commonly used features are [83] energy, entropy, signal-to-noise ratio (SNR), etc. The obtained features are used train the ANN for identification if the system in islanded or not. During training, different condition of operation (islanding as well as non-islanding) are simulated to create a database. Non islanding condition may include line to ground fault, line to line fault, disconnection of capacitive load, three-phase to ground fault and other various states of DGs. For the obtained database, about 75% of the data is used for the purpose of training, and 15–15% of data is used for testing and validation [84]. This method of islanding detection is found to have a high accuracy as per the literature.

6.4.3.2 FLC-based islanding detection technique

Fuzzy logic controller is one of the leading tool for system modeling in absence of specified mathematical formulation. During FLC the human knowledge is imitated by the system in terms of linguistic variables which are also known as fuzzy rules set [85]. For implementation of FLC-based control for islanding detection, input need to be determined. Variation in input value will determine the complexity of the FLC. In the literature [86], three parameter (voltage, change in frequency, utilized for islanding detection. The three parameters are monitored and based on the fuzzy rule set it is determined whether the system is in islanded on non-islanded condition. In another literature [87 inputs were considered. The increase in number of input make the FLC more accurate but at the same time the processing speed of the controller is reduced.

6.4.3.3 ANFIS-based islanding detection techniques

ANFIS is implemented for modeling non-linear system with less input and output data for training. It has ability to handle uncertainty like FLC and learning-based approach is present as in ANN. Because of advantages it is preferred over other techniques. Takagi-Sugeno method of FLC is utilized for ANFIS [88]. As per the literature [89], ANFIS-based islanding detection technique five input parameters (voltage, frequency, current, power and ratio of frequency and power) are considered. During the first stage, different condition are simulated to create a database for all the input condition response in different cases. In the later stages, the database is then used for training the ANFIS to evaluate the islanding condition efficiently. ANFIS-based islanding detection technique is easy to implement and can detect islanding condition accurately even for multilayer DGs system.

6.4.3.4 Decision tree-based islanding detection method

Decision tree is a pattern recognition-based classifier which provides solution for all the input based on the statistical variables. It help in solving the problem which cannot be solved easily by analytical methods. Fast learning capability of decision tree make it more suitable over the other pattern recognition techniques [90]. The decision tree breaks down the complex decision process into several decisions. The initial entry point is known as the root node which is then split into two or more child node depending upon the number of chooses available. A leaf node is formed where no more decision are to be made [91]. The decision tree explained above can be represented by Figure 6.15.

The method is widely implemented for islanding detection. In the literature [92], the voltage and current signal of the system is monitored, and DWT is used for extracting the features of the signal. The features are used to train the decision tree (DT) and random voltage and current signal is passed through the DT to classify if the system is operating in islanding condition or normal operating conditions.

Based on all the different intelligent islanding detection techniques discussed, in Table 6.6 different literatures are compared based on their accuracy of the techniques as presented in various literature.

6.5 Reactive power control and its effect on reliability

Based on the studies [96] it can be concluded that the electrical performance of the system is directly impacted by the thermal nature of the device leading to power loss and vice versa. The conduction and switching losses account for the major share in system power loss and even led to rise in temperature of the device caused because

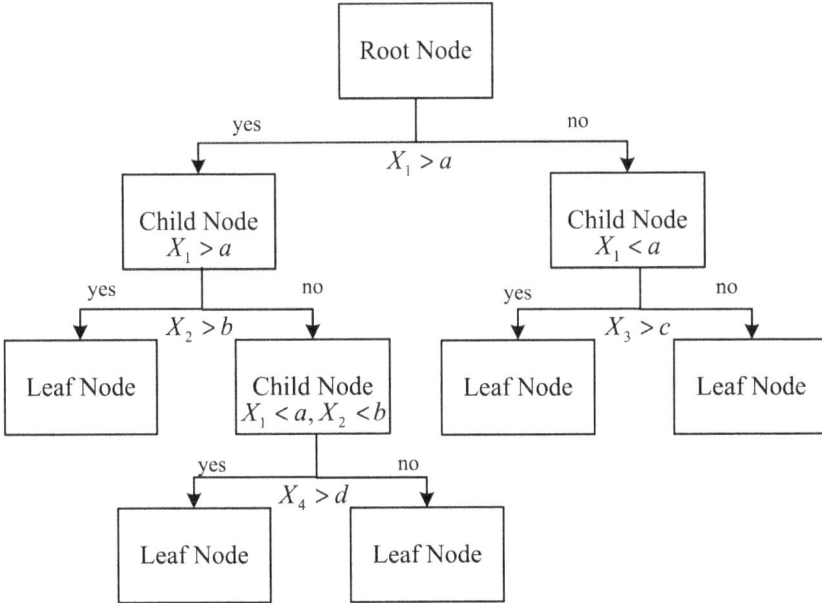

Figure 6.15   Decision tree schematic with leaf and child nodes [91]

of the thermal impedance present at the nodes. The power loss assists in studying the thermal stress on power electronic switches which may even cause device operation failure. Because of the above aspects, the reliability analysis prior to modeling of a power electronics converter is required [97, 98]. For power electronic application, it has been observed that the temperature variation and peak presents difficulty for reliability study [99, 100]. As a result, a stable junction is considered with mini-malistic variation in temperature for reliability analysis [98]. The coupling relation between junction temperature and power loss further helps in managing the device temperature by tuning the active and reactive power. The power control as discussed in Section 6.3, aid in regulating the junction temperature of a power electronics

Table 6.6   Comparative analysis of active islanding detection scheme

Islanding detection classifier	Signal processing techniques	Recognition rate (%)
ANN [93]	Wavelet transform	95
Decision tree [94]	DWT	94
Ridgelet probabilistic neural network [95]	DWT	93.3
Multi-feature-based ANN [84]	DWT	98.1

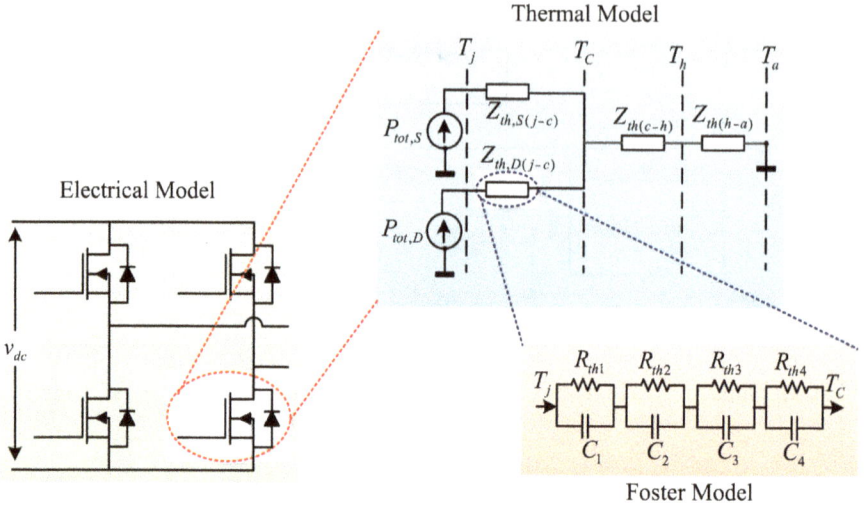

Figure 6.16   Schematic of the thermal model of the IGBT diode module

device. As a result, the power control strategies can enhance the reliability of the power electronic device and can further cause cost reduction in PV-based energy.

To further understand the thermal modeling of power electronic converter, a schematics representation is depicted in Figure 6.16. The modeling of insulated-gate bipolar transistor (IGBT) and the diode are done simultaneously because of the complementary nature of operating cycles. The power loss dissipated by the IGBT and diodes is represented by the thermal resistance $R_{th,S(j-c)}$ and $R_{th,D(j-c)}$, respectively. The thermal resistance lead to rise in junction temperature of IGBT(T_{Sj}) and diode (10kW) beyond ambient condition temperature (T_{Dj}). The foster model [101] is mostly used for the analysis and the thermal parameter required can be obtained from the device datasheet. The relation between the power loss and junction temperature, in case of both IGBT and diode, can be expressed by

$$T_j(t) = P_{tot}(t) Z_{th,(j-c)}(t) + T_c(t)$$ (6.51)

$$T_c(t) = T_a(t) + \left[P_{tot,S}(t) + P_{tot,D}(t) \right] \left[Z_{th(c-h)}(t) + Z_{th(h-a)}(t) \right]$$ (6.52)

where power loss of IGBT or diode is denoted by $P_{tot,S}$ and $P_{tot,D}$, respectively. The thermal impedance from junction to case are denoted by $Z_{th,(j-c)}$ and, thermal impedance from case to heatsink and heatsink to ambient condition are denoted by $Z_{th,(c-h)}$ and $Z_{th,(h-a)}$, respectively. The temperature of the case is represented by T_c. The steady state nature of junction temperature is mainly attributed to thermal resistance (C_{th}), whereas the dynamic behaviors is caused because of the thermal capacitance (C_{th}). The time constant for the thermal impedance is large which lead to slow dynamic response [102].

The relation between the temperature and power control is addressed through the flowchart presented in Figure 6.17. Power control helps us in achieving

$$T_j^*$$

$$\downarrow$$

Look-up table

$$\downarrow$$

Temp. determined
$(P_{j1}, Q_{j1}, P_{j2}, Q_{j2}, \ldots)$

$$\downarrow$$

Power reference $(P_1^*, Q_1^*, P_2^*, Q_2^*, \ldots)$	$\leftarrow$	Other objectives (P_L, Q_L)

$$\downarrow$$

Power optimization
$P^* = \max\{P_1^*, P_2^*, \ldots\}$

$$\downarrow$$

$$P^*, Q^*$$

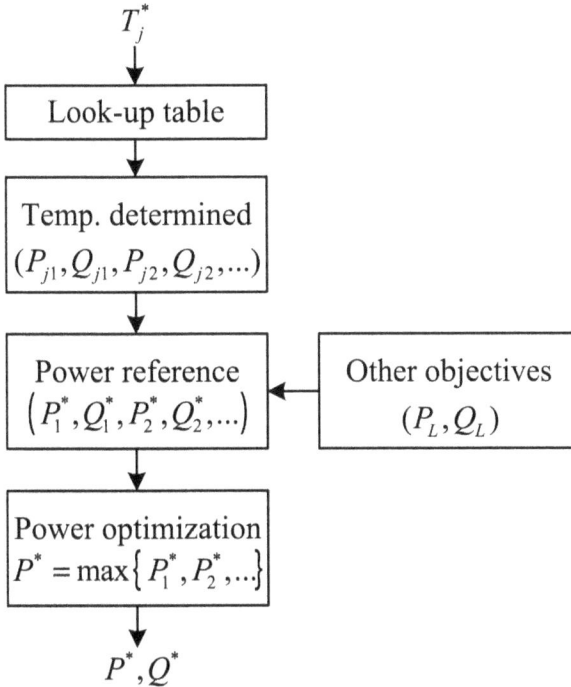

Figure 6.17   Flowchart for relating junction temperature with power control strategy

multi-objective, i.e., fault ride-he power reference $(P_1^*, Q_1^*, P_2^*, Q_2^* \ldots)$ need to be optimized for achieving the desired outcome.

In Figures 6.18 and 6.19, a simulated result of 10 kW single-phase and three-phase PV system are presented, respectively, to analyze the change in junction temperature with change in power control modes. It can be observed that by power control the IGBT junction temperature can remain constant even during the fault ride-through condition. And injection of sufficient reactive power helps the system to recover in time as well. As per the observation it can be concluded that multiple objectives are satisfied out of power control strategies.

6.6   Conclusion

The increase in installation of large capacity PV DG has reduced the pressure on the grid substantially, but at the same time has introduced many challenges during the interconnection process. The factors such as power quality, harmonics, flicker, etc., need to be regulated along with voltage and frequency level. In this chapter, the different regulation required for LVRT and anti-islanding have been discussed. And based on the regulation control strategies are presented for balancing the power and detecting the faulty condition.

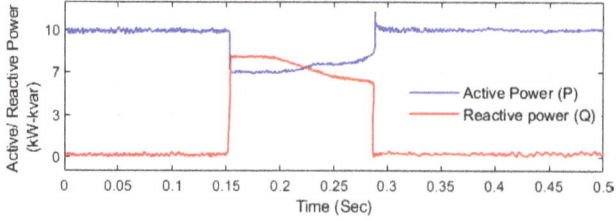

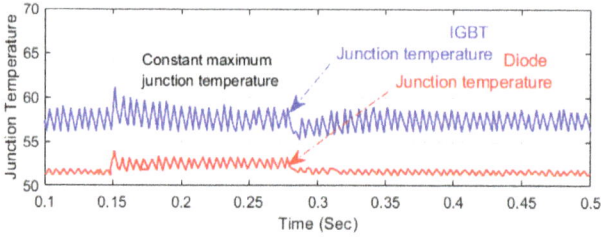

Figure 6.18   Simulation of 10 kW single-phase GCPVS with (a) active and reactive power, and (b) junction temperature for power device

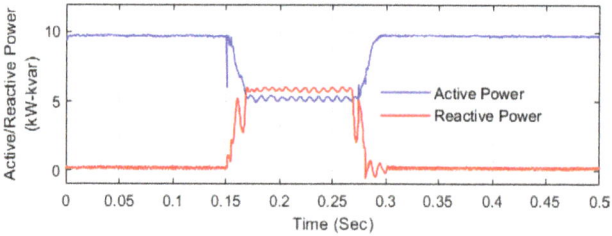

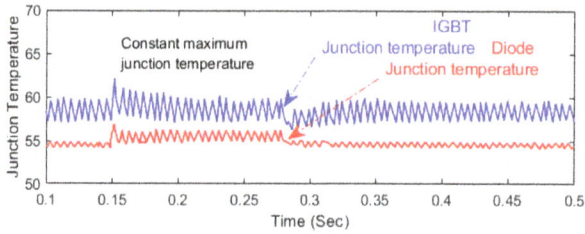

Figure 6.19   Simulation of 10 kW three-phase GCPVS with (a) active and reactive power and (b) junction temperature for power device

The aim of the chapter was to provide a brief regarding the requirement and enable the inverter controller with advance protection schemes.

References

[1] Yang Y., Blaabjerg F., Wang H. 'Low-voltage ride-through of single-phase transformerless photovoltaic inverters'. *IEEE Transactions on Industry Applications*. 2014;**50**(3):1942–52.

[2] Al-Shetwi A.Q., Sujod M.Z., Blaabjerg F., Yang Y. 'Fault ride-through control of grid-connected photovoltaic power plants: a review'. *Solar Energy*. 2019;**180**(2):340–50.

[3] Piya P., Ebrahimi M., Karimi-Ghartemani M., Khajehoddin S.A. 'Fault ride-through capability of voltage-controlled inverters'. *IEEE Transactions on Industrial Electronics*. 2018;**65**(10):7933–43.

[4] Kumar P., Singh A.K. (eds.) 'Grid codes: goals and challenges' in Hossain J., Mahmud A.. (eds.). *Renewable Energy Integration. Green Energy and Technology*. 1. **1**. Springer Science+Business Media Singapore; 2014.

[5] Office of Energy Projects. Energy policy act (EPAct) of 2005. Docket No. RM05-31-000 Order No. 665. Washington: Federal Energy Regulatory Commission; 2005. pp. 1–12.

[6] Hemus A. The grid code. Issue 5, Revision 21. National Grid House Warwick Technology Park Gallows Hill, Warwick CV34 6DA: National Grid Electricity Transmission plc; 2017. pp. 1–726.

[7] Grid Code Review Panel. EirGrid grid code. Version 3.3. Dublin, Ireland: Eirgrid Group; 2009. pp. 1–315.

[8] China Electric Power Research Institute. Technical rule for connecting wind farm to power system. GB/T 19963-2011. People's Republic of China: National Electricity Regulatory Standardization Technical Committee; 2011. pp. 1–12.

[9] Grid code high and extra high voltage. Bayreuth: E.ON Netz. GmbH; 2006. pp. 1–46. Available from http://www.pvupscale.org/IMG/pdf/D4_2_DE_annex_A-3_EON_HV_grid__connection_requirements_ENENARHS2006de. pdf.

[10] Electrical Installations Sectional Committee. National electrical code. ICS 01.120; 91.160.01. New Delhi: Goverment of India; 2011. pp. 1–379. Available from https://law.resource.org/pub/in/bis/S05/is.sp.30.2011.svg. html.

[11] Cronin T. *An overview of grid requirements in Denmark and the DTU advanced grid test facility at Østerild*. 2013DTU Wind Energy. Available from https://www. nrel.gov/grid/assets/pdfs/turbine_sim_13_denmark_grid_requirements.pdf.

[12] Standards N. Electromagnetic compatibility (EMC) part 3.2: Limits-Limits for harmonic current emissions (equipment input current less than or equal to 16 A per phase). EN IEC 61000-3-2:2019. Brussels: European Committee for Electrotechnical Standardization; 2019. pp. 1–18.

[13] Masoum M.A.S., Fuchs E.F. 'Introduction to Power Quality'. *in power quality Power Systems and Electrical Machines*. Elsevier; 2015. pp. 1–104.

[14] Poddar G., Ranganathan V.T. 'Direct torque and frequency control of double-inverter-fed slip-ring induction motor drive'. *IEEE Transactions on Industrial Electronics*. 2004;**51**(6):1329–37.

[15] Khan M.A., Haque A., Kurukuru V.S.B., Saad M. 'Advanced control strategy with voltage sag classification for single-phase grid-connected photovoltaic system'. *IEEE Journal of Emerging and Selected Topics in Industrial Electronics*. 2020:1–12.

[16] Fatama A.-Z., Khan M.A., Kurukuru V.S.B., Haque A. 'Hybrid algorithm for reactive power control in grid integrated photovoltaic inverters'. 2019 International Conference on Power Electronics, Control and Automation (ICPECA); 2019. pp. 1–6.

[17] Ahmad M. 'Power system frequency deviation measurement using an electronic bridge'. *IEEE Transactions on Instrumentation and Measurement*. 1988;**37**(1):147–8.

[18] Luo X., Wang J., Wojcik J., *et al.* 'Review of voltage and frequency grid code specifications for electrical energy storage applications'. *Energies*. 2018;**11**(5):1070.

[19] Mahmud N., Zahedi A., Mahmud A. 'A cooperative operation of novel PV inverter control scheme and storage energy management system based on ANFIS for voltage regulation of grid-tied PV system'. *IEEE Transactions on Industrial Informatics*. 2017;**13**(5):2657–68.

[20] Fatama A.Z., Khan M.A., Kurukuru V.S.B., Haque A., Blaabjerg F. 'Coordinated reactive power strategy using static synchronous compensator for photovoltaic inverters'. *International Transactions on Electrical Energy Systems*. 2020;**30**(6):1–18.

[21] Yang Y., Wang H., Blaabjerg F. 'Reactive power injection strategies for single-phase photovoltaic systems considering grid requirements'. *IEEE Transactions on Industry Applications*. 2014;**50**(6):4065–76.

[22] Zeng D., Wang G., Pan G., Li H. 'Fault ride-through capability enhancement of PV system with voltage support control strategy'. *Open Journal of Applied Sciences*. 2013;**03**(02):30–4.

[23] Khan M.A., Haque A., Kurukuru V.S.B. 'Droop based low voltage ride through implementation for grid integrated photovoltaic system'. 2019 International Conference on Power Electronics, Control and Automation (ICPECA); 2019. pp. 1–5.

[24] Radwan E., Nour M., Awada E., Baniyounes A. 'Fuzzy logic control for low-voltage ride-through single-phase grid-connected PV inverter'. *Energies*. 2019;**12**(24):4796.

[25] Vukosavic S.N. *Grid Side Converters Control and Design*. Springer US; 2018.

[26] Khan M.A., Haque A., Bharath V.S. 'Enhancement of fault ride through strategy for single-phase grid-connected photovoltaic systems'. 2019 IEEE Industry Applications Society Annual Meeting; 2019. pp. 1–6.

[27]   Tsili M., Papathanassiou S. 'A review of grid code technical requirements for wind farms'. *IET Renewable Power Generation*. 2009;**3**(3):308.

[28]   Tahir S., Wang J., Baloch M., Kaloi G. 'Digital control techniques based on voltage source inverters in renewable energy applications: a review'. *Electronics*. 2018;**7**(2):18.

[29]   Huang L., Xin H., Dörfler F. 'H-infinity-control of grid-connected converters: design, objectives and decentralized stability certificates'. *IEEE Transactions on Smart Grid*. 2020;**11**(5):1–8.

[30]   Aguilera R.P., Lezana P., Konstantinou G., *et al.*. Closed-loop SHE-PWM technique for power converters through model predictive control. IECON 2015 – 41st Annual Conference of the IEEE Industrial Electronics Society; 2015. pp. 005261–6.

[31]   Kouro S., Cortes P., Vargas R., Ammann U., Rodriguez J. 'Model predictive control—a simple and powerful method to control power converters'. *IEEE Transactions on Industrial Electronics*. 2009;**56**(6):1826–38.

[32]   Perez M.A., Fuentes R.L., Rodriguez J. 'Predictive control of DC-link voltage in an active-front-end rectifier'. 2011 IEEE International Symposium on Industrial Electronics; 2011. pp. 1811–16.

[33]   Rivera M., Yaramasu V., Rodriguez J., Wu B. 'Model predictive current control of two-level four-leg inverters—Part II: Experimental implementation and validation'. *IEEE Transactions on Power Electronics*. 2013;**28**(7):3469–78.

[34]   Hamidi A., Ahmadi A., Feali M.S., Karimi S. 'Implementation of digital FCS-MP controller for a three-phase inverter'. *Electrical Engineering*. 2015;**97**(1):25–34.

[35]   Rivera M., Yaramasu V., Llor A., Rodriguez J., Wu B., Fadel M. 'Digital predictive current control of a three-phase four-leg inverter predictive current control of a three-phase four-leg'. *IEEE Transactions on Industrial Electronics: A Publication of the IEEE Industrial Electronics Society*. 2013;**60**(11):4903–12.

[36]   Calle-Prado A., Alepuz S., Bordonau J., Nicolas-Apruzzese J., Cortes P., Rodriguez J. 'Model predictive current control of grid-connected neutral-point-clamped converters to meet low-voltage ride-through requirements'. *IEEE Transactions on Industrial Electronics*. 2015;**62**(3):1503–14.

[37]   Young H.A., Perez M.A., Rodriguez J., Abu-Rub H. 'Assessing finite-control-set model predictive control: a comparison with a linear current controller in two-level voltage source inverters'. *IEEE Industrial Electronics Magazine*. 2014;**8**(1):44–52.

[38]   Teodorescu R., Liserre M., Rodríguez P. *Grid Converters for Photovoltaic and Wind Power Systems*. John Wiley & Sons; 2011.

[39]   Singh P., Pradhan A.K. 'A local measurement based protection technique for distribution system with photovoltaic plants'. *IET Renewable Power Generation*. 2020;**14**(6):996–1003.

[40]   Suman M., Venkata Kirthiga M. 'Unintentional islanding detection'. *Distributed Energy Resources in Microgrids*. Elsevier; 2019. pp. 419–40.

[41] Salmi T., Bouzguenda M., Gastli A., Masmoudi A. 'Review of common-mode voltage in transformerless inverter topologies forf common-mode voltage in transformerless inverter topologies for PV systems'. *Smart Innovation, Systems and Technologies*. 2012;**12**:589–96.

[42] Khamis A., Shareef H., Bizkevelci E., Khatib T. 'A review of islanding detection techniques for renewable distributed generation systems'. *Renewable and Sustainable Energy Reviews*. 2013;**28**:483–93.

[43] IEEE. IEEE Std 1547, 2003. *IEEE Standard for Interconnecting Distributed Resources with Electric Power Systems*; 2003.

[44] IEEE. IEEE 1547,2008. *IEEE Standard for Interconnecting Distributed Resources with Electric Power Systems*; 2008.

[45] IEEE. IEEE Std 1547, 2014. *IEEE Standard for Interconnecting Distributed Resources with Electric Power Systems*; 2014.

[46] IEEE. IEEE P1547.1a. *IEEE Standard for Interconnecting Distributed Resources with Electric Power Systems*; 2014.

[47] IEEE-SA Standards Board. *IEEE Standard for Interconnection and Interoperability of Distributed Energy Resources with Associated Electric Power Systems Interfaces*. IEEE STD 1547-2018. Piscataway, New Jersey, United States: IEEE Standards association; 2018. pp. 1–116. 9781504449410.

[48] Solar photovoltaic energy systems. Utility-interconnected photovoltaic inverters - Test procedure of islanding prevention measures. IEC 62116-2014. Geneva, Switzerland: International Electrotechnical Commission; 2014. pp. 1–13.

[49] Richard De Blasio. IEEE recommended practice for utility interface of photovoltaic (pv) systems. IEEE 929-2000. Piscataway, New Jersey, United States: IEEE Standards Association; 2000. pp. 1–32. 9780738119359.

[50] Stadler I. Study about international standards for the connection of small distributed generators to the power grid. 2007.2189.4-001.00. GIZ Brazil: Deutsche Gesellschaft für Internationale Zusammenarbeit (GIZ) GmbH; 2011. pp. 1–65.

[51] Selbsttätige Schaltstelle zwischen einer netzparallelen Eigenerzeugungsanlage und dem öffentlichen Niederspannungsnetz. Automatic disconnection device between a generator and the public low-voltage grid. DIN VDE V 0126-1-1:2013-08;VDE V 0126-1-1:2013-08. Germany: Beuth Publishing DIN; 2005. pp. 1–10.

[52] Chowdhury S.P., Chowdhury S., Ten C.F., Crossley P.A. Islanding Protection of Distribution Systems with Distributed Generators – A Comprehensive Survey Report. 2008 IEEE Power and Energy Society General Meeting - Conversion and Delivery of Electrical Energy in the 21st Century; Pittsburgh, PA, USA, 20-24 July 2008; 2008. pp. 1–6.

[53] Ustun T.S., Hashimoto J., Otani K. 'Impact of smart inverters on feeder hosting capacity of distribution Networks'. *IEEE Access: Practical Innovations, Open Solutions*. 2019;7:163526–36.

[54] Xu W., Martel S., Mauch K. An assessment of distributed generation islanding detection methods and issues for Canada. TRN: CA0402591. Natural

Resources Canada, Varennes, PQ (Canada): Government of Canada, Ottawa, ON (Canada). Climate Change Action Fund; Natural Resources Canada, Ottawa, ON (Canada). Office of Energy Research and Development; 2004. pp. 1–53.

[55]   Laghari J.A., Mokhlis H., Karimi M., Bakar A.H.A., Mohamad H. 'Computational intelligence based techniques for islanding detection of distributed generation in distribution network: a review'. *Energy Conversion and Management*. 2014;**88**(NR 2a):139–52.

[56]   Laaksonen H. 'Advanced islanding detection functionality for future electricity distribution networks Is landing detection functionality for future electricity distribution networks'. *IEEE Transactions Power Delivery*. 2013;**28**(4):2056–64.

[57]   Xu W., Zhang G., Li C., Wang W., Wang G., Kliber J. 'A power line signaling based technique for anti-islanding protection of distributed generators—Part I: Scheme and analysis'. *IEEE Transactions on Power Delivery*. 2007;**22**(3):1758–66.

[58]   Riley C.M., Lin B.K., Habetler T.G., Kliman G.B. 'Stator current harmonics and their causal vibrations: a preliminary investigation of sensorless vibration monitoring applications'. *IEEE Transactions on Industry Applications*. 1999;**35**(1):94–9.

[59]   Sarangi S., Pradhan A.K. 'Adaptive direct underreaching transfer trip protection scheme for the three-terminal line'. *IEEE Transactions on Power Delivery*. 2015;**30**(6):2383–91.

[60]   Dutta S., Sadhu P.K., Jaya Bharata Reddy M., Mohanta D.K. 'Shifting of research trends in islanding detection method—a comprehensive survey'. *Protection and Control of Modern Power Systems*. 2018;**3**(1):1.

[61]   Ye Z., Kolwalkar A., Zhang Y., Du P., Walling R. 'Evaluation of anti-islanding schemes based on nondetection zone concept'. *IEEE Transactions on Power Electronics*. 2004;**19**(5):1171–6.

[62]   Li C., Cao C., Cao Y., Kuang Y., Zeng L., Fang B. 'A review of islanding detection methods for microgrid'. *Renewable and Sustainable Energy Reviews*. 2014;**35**(2):211–20.

[63]   Mahat P., Chen Z., Bak-Jensen B. 'Review of islanding detection methods for distributed generation'. 2008 Third International Conference on Electric Utility Deregulation and Restructuring and Power Technologies; 2008. pp. 2743–8.

[64]   Seyedi M., Taher S.A., Ganji B., Guerrero J.M. 'A hybrid islanding detection technique for inverter-based distributed generator units'. *International Transactions on Electrical Energy Systems*. 2019;**29**(11):1–21.

[65]   Hasna Mubarak Saleh Al Seiari. *An islanding detection method for micro-grids with grid-connected and islanded capability islanding detection method for micro-grids with grid-connected and islanded capability* scholarworks@UAEU. [Master of Science in Electrical Engineering (MSEE)] UAE, United Arab Emirates University. 2016. Available from https://scholarworks.uaeu.ac.ae/all_theses/463.

[66] Ciobotaru M., Agelidis V.G., Teodorescu R., Blaabjerg F. 'Accurate and less-disturbing active anti islanding method based on PLL for grid-connected con grid-connected converters'. *IEEE Transactions on Power Electronics*. 2010;**25**(6):1576–84.

[67] Lathi B.P., Ding Z. *Modern digital and analog communication systems*. 4. **1**. Oxford, United Kingdom: Oxford; 1998. pp. 1–928. 978-0198073802.

[68] Reddy C.R., Reddy K.H. 'A new passive islanding detection technique for integrated distributed generation system using rate of change of regulator voltage over reactive power at balanced islanding'. *Journal of Electrical Engineering & Technology*. 2019;**14**(2):527–34.

[69] Massoud A.M., Ahmed K.H., Finney S.J., Williams B.W. 'Harmonic distortion-based island detection technique for inverter-based distributed generation'. *IET Renewable Power Generation*. 2009;**3**(4):493.

[70] Merino J., Mendoza-Araya P., Venkataramanan G., Baysal M. 'Islanding detection in microgrids using harmonic signatures, detection in micro-grids using harmonic signatures'. *IEEE Transactions Power Delivery*. 2015;**30**(5):2102–9.

[71] Hossain J., Mahmud A. 'Fault Ride Through Criteria Challeges'. *Renewable energy integration: challenges and solutions*. 1. **1**. Singapore: Springer; 2014. pp. 1–447. 9814585270.

[72] Li W., Gu Y., Luo H., Cui W., He X., Xia C. 'Topology review and deriva-tion methodology of single-phase transformerless photovoltaic inverters for leakage current suppression'. *IEEE Transactions on Industrial Electronics*. 2015;**62**(7):4537–51.

[73] Shang Y., Shi S., Dong X. 'Islanding detection based on asymmetric tripping of feeder circuit breaker in ungrounded power distribution system'. *Journal of Modern Power Systems and Clean Energy*. 2015;**3**(4):526–32.

[74] Ropp M.E., Begovic M., Rohatgi A. 'Prevention of islanding in grid-connected photovoltaic systems'. *Progress in Photovoltaics: Research and Applications*. 1999;**7**(1):39–59.

[75] Reis M.V.G., Barros T.A.S., Moreira A.B., Nascimento F. E., Ruppert F., Villalva M.G. Analysis of the Sandia frequency shift (SFS) islanding detec-tion method with a single-phase photovoltaic distributed generation system. 2015 IEEE PES Innovative Smart Grid Technologies Latin America (ISGT LATAM); 2015. pp. 125–9.

[76] Wang X., Freitas W., Xu W., Dinavahi V. 'Impact of DG interface controls on the Sandia frequency shift antiislanding method'. *IEEE Transactions on Energy Conversion*. 2007;**22**(3):792–4.

[77] Ku Ahmad K.N.E., Selvaraj J., Rahim N.A. 'A review of the islanding detec-tion methods in grid-connected PV inverters'. *Renewable and Sustainable Energy Reviews*. 2013;**21**:756–66.

[78] Menon V., Nehrir M.H. 'A hybrid islanding detection technique using voltage unbalance and frequency set point'. *IEEE Transactions on Power Systems*. 2007;**22**(1):442–8.

[79] Elnozahy M., El-Saadany E., Salama M. 'A Robust Wavelet-ANN based technique for islanding detection'. Power and Energy Society General Meeting; 2011. pp. 1–8.

[80] Samet H., Hashemi F., Ghanbari T. 'Minimum non detection zone for islanding detection using an optimal artificial neural network algorithm based on PSO'. *Renewable and Sustainable Energy Reviews*. 2015;**52**(3):1–18.

[81] Fatama A., Haque A., Khan M.A. 'A Multi Feature Based Islanding Classification Technique for Distributed Generation Systems'. 2019 International Conference on Machine Learning, Big Data, Cloud and Parallel Computing (COMITCon); 2019. pp. 160–6.

[82] Khan M.A., Haque A., Kurukuru V.S.B. 'An efficient islanding classification technique for single phase grid connected photovoltaic system'. 2019 International Conference on Computer and Information Sciences (ICCIS); 2019. pp. 1–6.

[83] Kurukuru V.S.B., Blaabjerg F., Khan M.A., Haque A. 'A novel fault classification approach for photovoltaic systems'. *Energies*. 2020;**13**(2):308.

[84] Khan M.A., Kurukuru V.S.B., Haque A., Mekhilef S. 'Islanding classification mechanism for grid-connected photovoltaic systems'. *IEEE Journal of Emerging and Selected Topics in Power Electronics*. 2020:1.

[85] Khan M.A., Haque A., Kurukuru V.S.B. 'Intelligent control of a novel transformerless inverter topology for photovoltaic applications'. *Electrical Engineering*. 2020;**102**(2):627–41.

[86] Eugeniusz R., Burek A., Jedut L. A new method for islanding detection in distributed generation. 5th International Conference on Electrical and Electronics Engineering. Electric control. ELECO 2007; Bursa, Turkey, 5-9 December 2007; 2007. pp. 1–5.

[87] Samantaray S.R., El-Arroudi K., Joos G., Kamwa I. 'A fuzzy rule-based approach for islanding detection in distributed generation'. *IEEE Transactions on Power Delivery*. 2010;**25**(3):1427–33.

[88] Khan M.A., Haque A., Kurukuru V.S.B. 'Performance assessment of standalone transformerless inverters'. *International Transactions on Electrical Energy Systems*. 2020;**30**(1):1–21.

[89] Bitaraf H., Sheikholeslamzadeh M., Ranjbar A.M., Mozafari B. 'Neurofuzzy islanding detection in distributed generation'. IEEE PES Innovative Smart Grid Technologies; 2012. pp. 1–5.

[90] Khan M.A., Haque A., Kurukuru V.S.B. 'Machine learning based islanding detection for grid connected photovoltaic system'. 2019 International Conference on Power Electronics, Control and Automation (ICPECA); 2019. pp. 1–6.

[91] Gupta N., Garg R. 'Algorithm for islanding detection in photovoltaic generator network connected to low-voltage grid'. *IET Generation, Transmission & Distribution*. 2018;**12**(10):2280–7.

[92] Heidari M., Seifossadat G., Razaz M. 'Application of decision tree and discrete wavelet transform for an optimized intelligent-based islanding detection method in distributed systems with distributed generations'. *Renewable and Sustainable Energy Reviews*. 2013;**27**(4):525–32.

[93]   Fayyad Y. 'Neuro-wavelet based islanding detection technique'. IEEE Electrical Power & Energy Conference; 2010. pp. 1–6.

[94]   Parzen E. 'On estimation of a probability density function and mode in statistics'. *Annals of Mathematics*. 1962;**33**:1065–76.

[95]   Ahmadipour M., Hizam H., Othman M.L., Radzi M.A. 'Islanding detection method using ridgelet probabilistic neural network in distributed generation'. *Neurocomputing*. 2019;**329**(11):188–209.

[96]   Yang Y., Wang H., Blaabjerg F., Kerekes T. 'A hybrid power control concept for pv Inverters with reduced thermal loading'. *IEEE Transactions on Power Electronics*. 2014;**29**(12):6271–5.

[97]   Wang H., Liserre M., Blaabjerg F., *et al.* 'Transitioning to physics-of-failure as a reliability driver in power electronics'. *IEEE Journal of Emerging and Selected Topics in Power Electronics*. 2014;**2**(1):97–114.

[98]   Kurukuru V.S.B., Haque A., Khan M.A., Tripathy A.K. 'Reliability analysis of silicon carbide power modules in voltage source converters'. 2019 International Conference on Power Electronics, Control and Automation (ICPECA); 2019. pp. 1–6.

[99]   Brckner T., Bernet S. 'Estimation and measurement of junction temperatures in a three-level voltage source converter'. *IEEE Transactions on Power Electronics*. 2007;**22**(1):3–12.

[100] Shahzad M., Bharath K.V.S., Khan M.A., Haque A. 'Review on reliability of power electronic components in photovoltaic inverters'. 2019 International Conference on Power Electronics, Control and Automation (ICPECA); 2019. pp. 1–6.

[101] Anurag A., Yang Y., Blaabjerg F. 'Thermal performance and reliability analysis of single-phase PV inverters with reactive power injection outside Feed-In operating hours'. *IEEE Journal of Emerging and Selected Topics in Power Electronics*. 2015;**3**(4):870–80.

[102] Bryant A., Parker-Allotey N.-A., Hamilton D., *et al.* 'A fast loss and temperature simulation method for power converters, Part I: Electrothermal modeling and validation'. *IEEE Transactions on Power Electronics*. 2012;**27**(1):248–57.

Chapter 7

Thermal image based Monitoring of PV modules and Solar Inverters

Zainul Abdin Jaffery[1]

7.1 Introduction

As the concern for climate change is still increasing, the world is looking more toward clean energy resources and their efficient utilization. Solar energy has emerged as a major source of clean energy. In the last decades, solar energy installations have registered an exponential growth [1]. Amid the growing dependence on solar energy, reliability becomes an important issue in high-capacity grid-connected solar photovoltaic (PV) power systems. The two most important components of solar power systems are solar PV modules and inverters. Hence, to increase the reliability, efficiency, and safety of PV systems, fault detection and analysis of the two items become significant. Conventional fault-protection methods usually add fuses or circuit breakers in series with PV components. In general, the temperature of the components of electrical equipment is a common indicator of the health condition assessment. Loose connections, short circuits, or damaged components may cause an abnormally high temperature rise in electrical installations. Hence monitoring of temperature can be used effectively for the predictive fault analysis of PV systems. The infrared (IR) thermal imaging-based technique is a noninvasive method that analyzes the temperature of an electrical component or machinery [2]. IR thermal imaging has been successfully used in many applications [3, 4]. In this chapter, the use of IR thermal imaging has been discussed and used for fault analysis of solar PV system components.

7.2 Review of fault detection of PV modules

Many conventional methods of fault detection for solar PV modules have been proposed in the literature. Some of the widely used are PV characteristic measurement techniques, power loss analysis technique, arc detection analysis technique, ultrasonic inspection, and frequency analysis method. PV characteristic-based fault diagnosis methods [5] need a physical connection with the module to access the current

[1]Department of Electrical Engineering, Jamia Millia Islamia, India

and voltage output. This may interrupt the regular power supply and may be costly in terms of maintenance. In a large solar PV power plant, it is practically unfeasible to physically access every module. Other drawbacks of PV characteristic-based fault diagnosis methods include that they are time-consuming, especially in the case where a large number of PV modules at the plant level are required to be diagnosed.

A frequency analysis-based fault detection method under dark conditions using red light on PV array was proposed in [6]. The presence of an open-circuit fault in a PV array operated under dark conditions with a forward bias voltage produces an increase in the total dynamic impedance. As a result, the spectral components of the output voltage change. By analyzing the spectral components of the PV array, output voltage fault may be analyzed. However, this method can detect only the open-circuit fault in a PV array.

A power loss analysis method was used in [7] to find the faults in PV modules by comparing the simulated power with actual output power. However, one never be sure about the reason for the loss of power. Time-domain reflector and earth capacitor measurement have been used by researchers in [8, 9]. An arc detection-based fault diagnosing technique using compressed sensing has also been reported in [10]. This work adopted the high-frequency analysis using the analog-to-digital converter technology.

Although the above-discussed methods are well established, there are some limitations associated with these methods. The abnormal voltage–current characteristic curve of a PV panel is a clear indication of a fault, but to find the source or physical location of the fault (especially when a large number of modules are installed), this method may not be suitable. Hence to detect the exact location and nature of the fault, some other method is required. Additionally, the above methods are invasive techniques, which means that PV panels are needed to be physically connected to some measuring devices. Hence there is always some risk of interfering with the normal operation of PV modules.

Recently IR image-processing-based techniques for the condition monitoring of machinery are becoming popular in industries [2]. The use of IR imaging for PV plant inspection offers several advantages. Anomalies can be seen on a crisp thermal image. The exact location of faults can be detected. IR imaging is a noninvasive technique; hence, PV panels can be inspected without disturbing the normal operation. A large number of panels can be scanned in a short time duration using thermal imaging cameras. Moreover, IR imaging offers early fault-diagnosis before the complete failure of the system. Hence, to ensure failure-free operation, IR imaging is a simple, fast, and reliable method to scan the faults in solar panels.

7.3   Thermal image-processing-based fault analysis

Every object emits IR radiations above the absolute zero temperature. The intensity of emitted radiation is proportional to the temperature of the object. For correct measurement, ambient temperature values and emissivity of the material needs to be adjusted accurately in the camera. For example, the emissivity of glass is 0.85 and for the polymer back-sheet, it is 0.95. Thermal cameras capture the temperature anomaly in the form of IR images. This temperature difference arises due to the various defects on the surface of an object.

In PV modules, various types of hot-spot patterns emerge due to different types of faults. For example, one string of a module may be warmer than others as shown in Table 7.1. This may be due to short-circuited or open sub-string. It will result in power losses or reduced open-circuit voltage. In other cases, the temperature of one cell is much larger than the others. This may be due to broken cells, disconnected strings, or diode burnt. Major IR image patterns are given in Table 7.1 along with the possible causes and their effects on the PV power output.

Thermal imaging has already been suggested in the literature for preventive fault diagnosis in PV modules. Recent research developments in the diagnosis of PV crystalline modules using thermal imaging were highlighted in [14]. A hot-spot analysis was done in [11] by analyzing the line temperature profile across the hot spots. In this technique, the temperature line profile of a healthy module is compared to the line profile of the PV module under testing.

As discussed previously, feature extraction is an important task in fault diagnosis using thermal imaging-based techniques. Various temperature-based features such as minimum temperature, maximum temperature, and background temperature can be extracted from thermal images, and then a decision-based system is employed to find the types of faults as discussed in [12]. Digital image-processing techniques have been used on thermal images to find the faults in PV modules. Canny edge detection is the most suitable method used in fault detection in thermal images of solar PV modules [13].

A process flow diagram for the fault analysis of solar PV modules using thermal image processing is given in Figure 7.1.

7.3.1   Preprocessing

Thermal images are inherently prone to noise. IR detectors are very sensitive to ambient temperature conditions. They easily capture the IR rays emitted from other objects that lie nearby. Different types of noises may be present in images such as Gaussian noise, impulse noise, shot noise, and speckle noise. To have accurate information from IR images, it is necessary to enhance signal content and suppress noise to increase the signal-to-noise ratio. It is very difficult to locate the faulty regions in images due to minor intensity differences between faulty regions and background regions. Therefore, preprocessing is essential to enhance the intensity difference between the targeted region and the background region without altering the important features of an image. Image preprocessing consists of two steps, namely, contrast enhancement and de-noising. In image de-noising, residual noise in the IR image is filtered out with an appropriate kind of filter. Fixed-pattern noise consists of high spatial frequencies and can be removed by low-pass filters, which suppress high frequencies and, at the same time, preserve edges and contours [15, 16].

Table 7.1 Frequently occurring fault categories in PV modules with their causes and effects

IR image	Fault description	Possible failure reason	Remarks
	String fault [11]	Open or short sub-string, bypass diode short circuit	Reduced power output, less V_{oc}
	Single-cell burnt, diode burnt, defected cell [12]	Physical damage, diode failure, shadowing	Reduced power output, physical inspection required
	Dust on PV module [13]	Dust accumulated on PV module	Reduced power output (temporarily)
	Manufacturing defect	Due to diode mismatch while assembling, nonuniform materials used	Increased temperature of the module

(Continues)

Table 7.1 *Continued*

IR image	Fault description	Possible failure reason	Remarks
	Healthy module	No defect	No defect
	Patchwork pattern	Bird dropping, shadowing	Decreased power output, overheating
	Physical damage discoloration	Damage due to mishandling during transportation or installation	Overheating of the panel, decreased power output

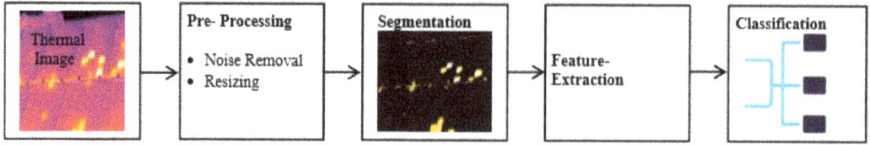

Figure 7.1   Fault analysis of solar PV modules using thermal image processing

7.3.2   Image segmentation

Image segmentation is a process to find the region of interest (ROI) in the thermal images. In most commercial applications, predefined geometrical shapes such as rectangles or circles are used for detection. The limitation with these predefined shapes is that sometimes they include the areas that are not necessary to be included in the ROI and sometimes they leave the area that is required to be included in the ROI. To overcome this problem, automatic segmentation methods have been proposed in [17]. Segmentation algorithms are generally based on either discontinuity in the image region or the similarity of the pixels. In the first case, the approach is to partition an image based on abrupt changes, whereas in the second approach, the image is partitioned into regions that are similar as per predefined criteria. Several segmentation methods are available in the literature; some of the popular methods are thresholding method, region growing, watershed algorithm, and clustering method [18].

7.3.3   Feature extraction

Feature extraction and classification is an important part of thermal IR imaging applications. Both the classification and recognition algorithm and the selection of features, through their interaction, can affect the accuracy of the classification and recognition of targets. As one of the much-debated challenges, this issue has received a good deal of attention and research. Common shape-based features such as area, centroid, edge pixels, and Harris corner are taken for visual images [19]. The feature extraction method purely depends on the purpose of the thermal imaging application.

7.3.4   Image classification

Image classification is the process of assigning pixels in the image into different classes of interest. Inspired by the human visual system, many attempts have been made to replicate this process with computer systems. To enable computers to understand objects, it is necessary to create a system that would extract high-level features from visual "stimuli" using only numerical manipulations. For a long time, image classification was not considered as a statistical problem until the development of Artificial Neural Networks (ANNs), particularly, Convolutional Neural Networks (CNNs). CNN is a special type of ANN that works in the same way as a regular neural network except that it has a convolution layer at the beginning [20].

7.4 CNN-based fault diagnosis of solar PV modules

To overcome the drawbacks of traditional fault-analysis systems for solar PV modules, a thermal imaging-based fault-analysis system is proposed. A thermal inspection of PV plants can be done using three methods.

1. thermal inspection using handheld cameras
2. thermal inspection using IR cameras mounted on the crane
3. thermal inspection using drone-mounted IR cameras.

Handheld cameras are generally used to inspect stand-alone PV power systems or small-capacity roof-mounted PV modules. Crane-mounted IR cameras are used for large power plants up to 1 MW capacity. However, for very large power plants, manual inspection is costly and time-consuming. In such cases, IR cameras mounted on drones are most viable economically and time-saving [21].

Deep learning has emerged as a powerful machine learning tool in recent times. It is a biologically inspired model that simulates the human visual cortex that can extract important semantic information from an image or a region. Deep learning networks have found applications in almost every area and especially data mining and image classification [22]. A deep learning network consists of a large number of hidden layers. Although a simple neural network also contains multiple layers in the deep learning network, hidden layers are much larger than a simple neural network. CNN is the most common network used in deep learning networks. For training the CNNs, a large number of images are needed. The general architecture of CNN is shown in Figure 7.2. It consists of four layers: convolution layer, rectified linear unit (ReLU), pooling, and fully connected (FC) layer.

7.4.1 Convolution layer

In the convolution layer, a matrix known as the kernel is passed through the input matrix to create a feature map for the next matrix. Kernel size can be varied to calculate another set of features. Convolution filter maps a set of points to a single point

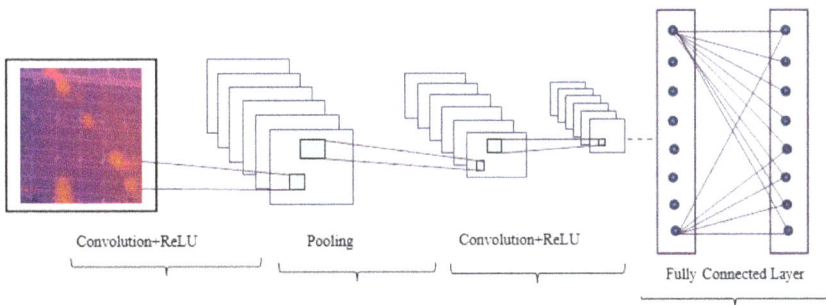

Convolution+ReLU   Pooling   Convolution+ReLU

Fully Connected Layer

Figure 7.2   General architecture of CNN

in the next layer. It takes the sum of the element-wise product of the filter and a part of the input matrix.

7.4.2   Rectified linear unit

ReLU is a piecewise linear function that will output the input directly if it is positive; otherwise it will give a zero. ReLU has become a default activation function for many neural networks because it gives better results and it is easier to train.

7.4.3   Pooling

While using the convolution filter on input images, a large number of small images are generated. To collect these small images into one single image, we use a pooling layer. Pooling takes the maximum of the filtered output from all these small matrices.

In this chapter, images extracted from a video of the aerial survey of a large PV power plant are fed to a deep convolution neural network. The input image layer is built with a dataset of 3 072 images, where each image is of size 393 × 424 × 3. Here 3 is the channel size and is used for color images. For grayscale images, the channel size is taken as 1. In convolution layers, the kernel size is taken as 3 × 3 and the number of filters taken per layer is 8. This parameter determines the number of feature maps.

Convolution layers are followed by a down-sampling operation. The down-sampling operation makes it possible to accommodate more convolution filters in deeper stages and hence increases the accuracy of the network. There are many methods of down-sampling. Max pooling [20] is one way of down-sampling, which is taken in this paper. Max pooling returns the maximum value of the rectangular regions of the inputs.

The number of fully connected layers gives the total number of categories in which the input images are to be classified. In the present work, seven fully connected layers are taken to classify the images according to the fault categories given in Table 7.1.

For different deep learning models and classification tasks, choosing the appropriate optimization algorithm plays an important role in improving the training speed and classification accuracy of the model. In this chapter, the three most commonly used optimization algorithms including adam (Adaptive Moment Estimation), rmsprop (Root Mean Square Propagation), and sgdm (Stochastic Gradient Descent with Momentum) are tested and compared as shown in the training graph for the deep learning network in Figure 7.3.

A deep learning net was tested for 674 images that included faulty as well as non-faulty images. The adam training network solver provided the best results among the three solvers taken. The confusion matrix for adam is shown in Figure 7.4. It provides an overall accuracy of 93.76 percent as compared to 88.5 percent and 91.69 percent in the case of rmsprop and sgdm, respectively, as shown in Table 7.2.

The performance of all the three training algorithms in terms of training accuracy, training time, and testing time are summarized in Table 7.2. It is concluded

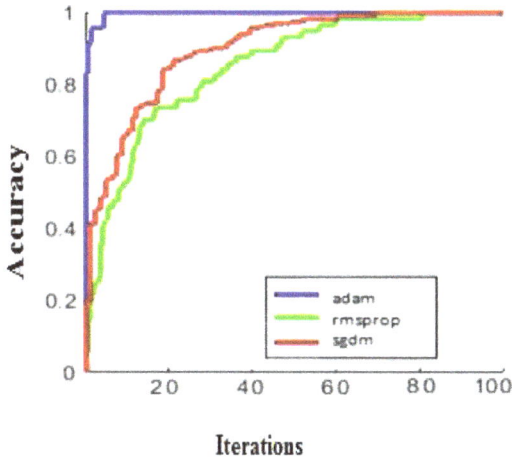

Figure 7.3   Training progress curve for adam, rmsprop, and sgdm

from the results that adam is the most accurate as well as fast training-solver among the three. It takes 115 seconds to train the deep learning net and 0.171 to test 674 images, whereas rmsprop is the least accurate among the three training solvers.

Results of fault category-wise correct and incorrect classification for the three training methods are given in Table 7.3. In the case of adam, 118 out of 674 PV modules were found non-faulty. A total of 138 modules were found faulty due to the patchwork pattern, which is a common cause of faults in PV modules due to

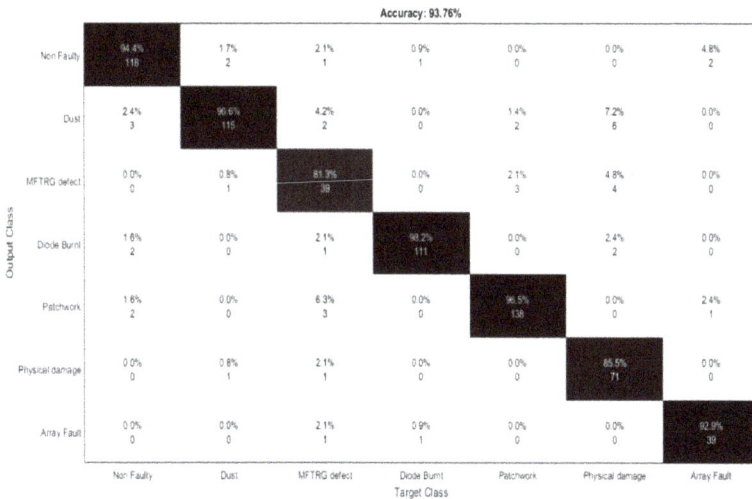

Figure 7.4   Confusion matrix for adam algorithm

Table 7.2   *Performance of training algorithms for the proposed deep net*

Training algorithm	Number of iterations	Accuracy (%)	Training time (s)	Testing time (s)
adam	100	93.76	115	0.171
rmsprop	100	88.50	118	0.184
sgdm	100	91.69	114	0.176

bird-droppings. From the results, it is also inferred that dust is also a major cause of the health degradation of panels.

7.5   CNN-based fault diagnosis of solar PV modules

Fault analysis of solar inverters is generally based on the electrical parameter's measurement and its analysis. Fault detection based on the current trajectory and its instantaneous frequency at the output side of PV inverters was done in [23]. ANN-based multilevel inverter fault detection system was proposed in [24], which delivers the power even when the fault occurs. A transformer-based multilevel converter was proposed in [25, 26]. A research facility design was proposed in [27] to detect inverter faults. Most of the research in PV inverter fault detection has been done to bypass the faulty components. Multilevel inverter strategies are suggested to make the system fault-tolerant. Research on pre-fault condition monitoring is rare in the literature. Besides, these techniques are invasive. In this chapter, a thermal imaging-based PV inverter condition-monitoring method is proposed, which diagnoses the thermal health of the inverter so that preventive actions can be taken before the complete system fails. The proposed method uses fuzzy decision-making to classify the health status of the transformer. The decision is made based on the maximum temperature of the components of the inverter and comparison of images of a non-faulty inverter with a faulty inverter. Thermal imaging is a noninvasive method that helps in quickly diagnosing the thermal health of inverter components. Thermal image processing is used to analyze the change in temperature distribution over the surface of the components and possible faults are predicted based on failure mode effect analysis (FMEA). Condition abnormalities due to aging, overloading, corrosion, and ambient temperature can be detected with this technique very easily and accurately.

7.5.1   Feature extraction

In the proposed technique, thermal images of the inverter under monitoring are taken with an IR camera. These images are quantitatively analyzed with the Smart View software, which gives two features of the image as similarity index and maximum temperature as shown in Figure 7.5. The inverter IR image is divided into zones. Each zone contains major components of the inverter as shown in Figure 7.6. For each zone, the maximum temperature is determined.

Table 7.3   Category-wise fault classification for deep learning net

| | | | | | Training accuracy | | | |
| | | adam | | rmsprop | | sgdm | | |
Fault category	Total images	Correct	Incorrect	Correct	Incorrect	Correct	Incorrect
Non-faulty	124	118	6	115	9	117	7
Dust	128	115	13	108	20	112	16
Manufacturing defect	47	39	8	36	11	36	11
Diode burnt	116	108	8	110	6	106	10
Patchwork	144	138	6	126	18	137	7
Physical damage	73	71	2	68	5	70	3
Array fault	42	39	3	34	8	40	2

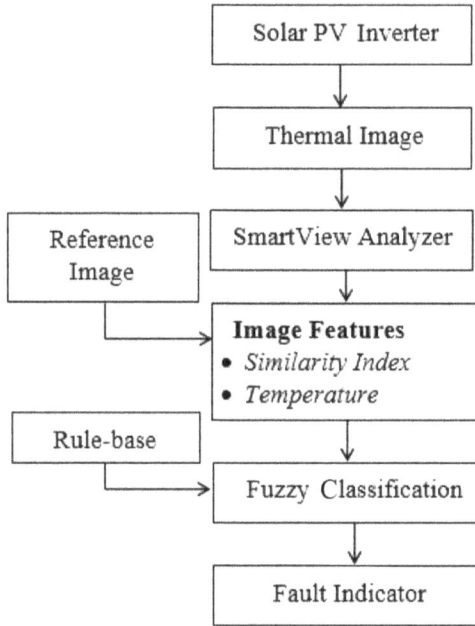

Figure 7.5   Condition monitoring algorithm for inverter

The manufacturer provides the maximum and minimum temperature rating of each component. Fault severity is decided based on the temperature of each component and the similarity index. The similarity index is calculated by taking the correlation between the IR images of the inverter under monitoring and the IR image of the same inverter when it is working in perfect conditions as

$$C\left(m,n\right) = \sum_{x=0}^{i-1}\sum_{y=0}^{j-1} f\left(x,y\right)\ g\left(x+m,\ y+n\right) \tag{7.1}$$

where $C\left(m,n\right)$ is the correlation between images $f\left(x,y\right)$ and $g\left(x,\ y\right)$.

If-then rule is designed using the similarity index and the temperature corresponding to the components of inverter. The IR image of a solar inverter is given in Figure 7.6.

A list of all major components of the inverter is prepared and the maximum temperature is determined for each component. The temperature range and similarity index are divided into low, medium, and high. The range of similarity index varies between 0 and 1 000. A fuzzy rule base is designed as given in Table 7.4.

Fault due to temperature rise of major components of an inverter is given in Table 7.5. Based on the FMEA, if the temperature goes out of the range given in Table 7.5, possible failure reasons are given. Maintenance actions are to be taken based on the severity of the fault. Critical faults are needed to be rectified immediately.

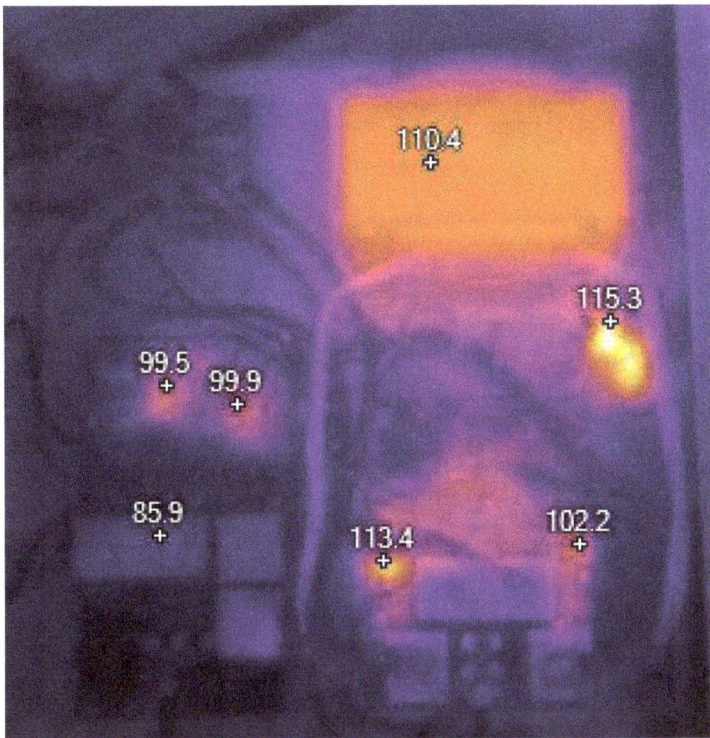

Figure 7.6   IR image of a solar inverter

The proposed technique helps to maintain the health of the solar PV inverter without interfering in the inverter operation, i.e., the system does not need to be shut down. This technique may also be used for quality testing at the inverter manufacturing unit before the product is sent to the market.

Table 7.4   Fuzzy rule base

	Similarity index			Temperature (T)		Fault category
if	LOW	and if	LOW	then	No fault
if	LOW	and if	MED	then	Minor fault
if	LOW	and if	HIGH	then	Major fault
if	MED	and if	LOW	then	Minor fault
if	MED	and if	MED	then	Major fault
if	MED	and if	HIGH	then	Major fault
if	HIGH	and if	LOW	then	Major fault
if	HIGH	and if	MED	then	Critical fault
if	HIGH	and if	HIGH	then	Critical fault

Table 7.5   Fault prediction based on FMEA

Important components	Manufacturer temperature range (°C)		Failure impact
	Maximum	Minimum	
STP80NF10	−65	170	Fault in DC–DC converter switching
LVC541A	−40	85	DC–DC converter driver failure
Inductor	−40	85	Smoothing fault
Capacitor	−55	105	Charging and discharging rate fault
Heatsink	−55	150	Inductor and capacitor burnout
BD-135	−55	150	DC–AC switching fault
Current sensor	−25	100	Problem in fault detection
Heatsink	−55	150	Switch burnout
Winding of transformer	0	115	Transformer malfunction or coil burnout
Heatsink	−55	150	Disconnection of battery half
Resistance	−55	155	Overcurrent flow in system
Fuse	−55	125	Fault left undetected

7.6 Summary

To increase the reliability of solar PV systems, condition monitoring of PV modules and inverters is needed without interfering the operation of the system. Thermal image-processing-based techniques are noninvasive and simple and can be used remotely with communication technologies. In this chapter, simple algorithms have been demonstrated using intelligent techniques such as CNN and fuzzy logic to analyze the conditions of PV modules and inverters. Simulation results illustrate the efficacy of these algorithms. Since the thermal images are taken in harsh environmental conditions, the reliability of the proposed methods depends on the selection of detection and classification algorithms.

References

[1]   Burduhos B.-G., Visa I., Duta A., Neagoe M. 'Analysis of the conversion efficiency of five types of photovoltaic modules during high relative humidity time periods'. *IEEE Journal of Photovoltaics*. 2018;**8**(No.6):1716–24.
[2]   Bagavathiappan S., Lahiri B.B., Saravanan T., Philip J., Jayakumar T. 'Infrared thermography for condition monitoring – a review'. *Infrared Physics & Technology*. 2013;**60**:35–55.

[3] Jadin M.S., Taib S. 'Recent progress in diagnosing the reliability of electrical equipment by using infrared thermography'. *Infrared Physics & Technology*. 2012;**55**(4):236–45.

[4] Jaffery Z.A., Dubey A.K. 'Design of early fault detection technique for electrical assets using infrared thermograms'. *International Journal of Electrical Power & Energy Systems*. 2014;**63**(9):753–9.

[5] Tingting P., Xiaohong H. 'A fault detection method for photovoltaic systems based on voltage and current observation and evaluation'. *Energies (MDPI)*. 2019;**12**(9):1–16.

[6] Perla Y. 'A novel fault detection and location method for PV arrays based on frequency analysis'. *IEEE Access :Practical Innovations, Open Solutions*. 2019;**7**:2169–3536.

[7] Chen Y.H., Liang R., Tian Y., Wang F. 'A novel fault diagnosis method of PV based-on power loss and I-V characteristics'. International Conference on New Energy and Future Energy System (NEFES 2016); Beijing, China; August 2016.

[8] Takashima T., Yamaguchi J., Otani K., Kato K., Ishida M. 'Experimental studies of failure detection methods in PV module strings'. Photovoltaic Energy Conversion, Conference Record of the IEEE 4th World Conference, 2; 2006. pp. 2227–30.

[9] Takashima T., Yamaguchi J., Otani K., Oozeki T., Kato K., Ishida M. 'Experimental studies of fault location in PV module strings'. *Solar Energy Materials and Solar Cells*. 2009;**93**(6):1079–82.

[10] Fenz W., Thumfart S., Yatchak R., Roitner H., Hofer B., *et al.* 'Detection of arc faults in PV systems using compressed sensing'. *IEEE Journal of Photovoltaics*. 2020;**10**(2):676–84.

[11] Simon M., Meyer E.L. 'Detection and analysis of hot-spot formation in solar cells'. *Solar Energy Materials and Solar Cells*. 2010;**94**(2):106–13.

[12] Jaffery Z.A., Dubey A.K., Irshad., Haque A. 'Scheme for predictive fault diagnosis in photo-voltaic modules using thermal imaging'. *Infrared Physics & Technology*. 2017;**83**(12):182–7.

[13] Tsanakas J.A., Chrysostomou D., Botsaris P.N., Gasteratos A. 'Fault diagnosis of photovoltaic modules through image processing and Canny edge detection on field thermographic measurements'. *International Journal of Sustainable Energy*. 2015;**34**(6):351–72.

[14] Kumar S., Jena P., Sinha A., Gupta R. 'Application of infrared thermography for non-destructive inspection of solar photovoltaic module'. *Journal of Non-Destructive Test Evaluation*. 2017;**15**(9):25–32.

[15] Buzdan R., Wyzgolik R. 'Remarks on noise removal in infrared images'. *Measurement Automation Monitoring*. 2015;**61**(no. 06).

[16] Vries J.D. 'Image processing and noise reduction techniques for thermographic images from large-scale industrial fires'. *Quantitative Infrared Thermography*. 2014;**042**:1–9.

[17] Irshad; Jaffery, Z.A. 'Performance comparison of image segmentation techniques for infrared images'. 2015 Annual IEEE India Conference (INDICON), New Delhi; 2015. pp. 1–5.

[18] Jaffery Z.A., Zaheeruddin S.L. 'Performance analysis of image segmentation methods for the detection of masses in mammograms'. *International Journal of Computer Applications*. 2013;**82**(2):44–50.

[19] Lowe D.G. 'Distinctive image features from scale-invariant keypoints'. *International Journal of Computer Vision*. 2004;**60**(2):91–110.

[20] Shin H.-C., Roth H.R., Gao M., *et al.* 'Deep convolutional neural networks for computer-aided detection: CNN architectures, dataset characteristics and transfer learning'. *IEEE Transactions on Medical Imaging*. 2016;**35**(5):1285–98.

[21] Yahya Z., Achraf E., Imane S., Sara A.L. 'Thermal infrared and visual inspection of photovoltaic installations by UAV photogrammetry—application case: Morocco'. *Drones*. 2018;**2**(41).

[22] Xin M., Wang Y, Yong W. 'Research on image classification model based on deep convolution neural network'. *EURASIP Journal on Image and Video Processing*. 2019;**2019**(1).

[23] Guan Y., Sun D., He Y. 'Mean current vector based online real-time fault diagnosis for voltage source inverter fed induction motor drives'. IEEE International Electric Machines & Drives Conference, 2; 2006. pp. 1114–18.

[24] Stonier A.A., Lehman B. 'An intelligent-based fault-tolerant system for solar-fed cascaded multilevel inverters'. *IEEE Transactions on Energy Conversion*. 2018;**33**(3):1047–57.

[25] Nademi H., Das A., Burgos R., Norum L.E., Lars E.N. 'A new circuit performance of modular multilevel inverter suitable for photovoltaic conversion plants'. *IEEE Journal of Emerging and Selected Topics in Power Electronics*. 2016;**4**(2):393–404.

[26] Kim S.-M., Lee J.-S., Lee K.-B. 'A modified level-shifted PWM strategy for fault-tolerant cascaded multilevel Inverters with improved power distribution'. *IEEE Transactions on Industrial Electronics*. 2016;**63**(11):7264–74.

[27] Rao S. 'An 18 kW solar array research facility for fault detection experiments'. Proceedings of the 18th Mediterranean Electro Technical Conference; Limassol, Cyprus; 2016.

Chapter 8

Failure mode classification for grid-connected photovoltaic converters

V S Bharath Kurukuru[1], Mohammed Ali Khan[1], and Azra Malik[1]

Anomaly detection with machine learning (ML) techniques has been extensively used to assist the decision-making process during abnormal conditions. However, these approaches were majorly constrained in the fields of medical and image processing-based applications. This is because of the complexities due to the unavailability and uncertainty of the input data of real-world and mainly industrial applications. In light of the above observation, this chapter proposes a failure mode classification (FMC) for power electronics converters (PECs) that are associated with grid-connected photovoltaic (PV) applications. Generally, the failure mechanisms are a critical aspect of determining the reliability of power electronic converters. The failure mechanisms deal with the physical, chemical and electrical processes through which the failure occurs in the system. Based on the type of failure process, these failure mechanisms can be modelled when appropriate material and environmental information are available. Moreover, with the advancements in ML approaches, the failure data along with the modelled mechanisms can be used to identify the operating state of the PEC. This helps to monitor the operation of the system, performing risk analysis, estimating the reliability of the product and reducing the probability that a customer is exposed to a potential product and or process problem. Further, the development of an FMC for a PEC can be discussed as follows: Initially, all the possible failure modes for PEC components can be identified. Further, the identified data is represented as a function of time and frequency domain to perform the feature extraction process. Further, the extracted feature data is minimized and subjected to classification using ML techniques. Finally, the trained ML classifier is implemented in a feed-forward loop to identify the operating state of a PEC.

[1]Advance Power Electronics Research Lab, Department of Electrical Engineering, Jamia Millia Islamia University, New Delhi, India

8.1   Introduction

With the immense expansion in the global population rate, the proportionate growth in the global energy demand is inevitable. There is an immediate need to incorporate renewable energy solutions to the existing energy infrastructure [1]. During the past few years, these renewable energy solutions are being preferred to reduce the stress on the traditional energy resources. Among the available renewable energy resources such as solar PV, wind energy, hydro, tidal, biomass, PV is considered to have highest generation capacity. To provide sustainable energy solution, grid-connected PV systems are increasingly utilized worldwide due to their inherent advantages. They help in reducing greenhouse gas emissions and provide clean and green energy. The reduction in the PV installation and maintenance cost along with the advanced and innovative power electronics technology has taken the grid-connected PV systems to a higher level. The grid-connected PV systems utilize power electronics converters for efficient power conversion. The switching devices based on semiconductor technology serve as the building blocks of power electronic converters [2]. These power devices are utilized for the conversion of current and voltage waveforms. These switching devices provide many benefits such as packaging size, high switching speed and flexible control. The conversion may include the increase or decrease in the current and voltage levels of the input or it may include the power conversion from DC to AC or vice versa. The power conversion should take place such that the system stability and reliability is maintained [3]. For grid-connected PV applications, inverters and converters are commonly used power electronic systems. An inverter acts as an interface between PV and the grid supply and it is responsible for the transfer of power from the DC side to AC side. These converters provide ancillary services such as voltage regulation and reactive power support at the point of grid connection.

The grid-connected PV systems are considered highly reliable. However, like any other complex electrical system, they are also susceptible to failures. These PV systems are modular in nature. Their capacity may vary from a few watts to megawatts. Thus, they can have different types of topologies and configurations [4], which further contributes to complicating the evaluation of failure modes of the system. Researchers have found that the vast majority of grid-connected PV system failures are inverter related. Inverters essentially are much different in behaviour from the rotating machines such as induction and synchronous machines. Their fault currents decay with a fast rate because they are deficient in inductive characteristics and their characteristics mainly depend upon the time constants of the circuit [5]. However, they can be controlled in such a way that their fault response time can be varied programmatically. Inverters can utilize voltage or control algorithm as the response to fault occurrence. During the transient period, the fault contribution is higher if the voltage control algorithm is used. With the current control algorithm, the increment or decrement rate is much lower. Inverters predominantly consist of the switching devices such as IGBT (insulated gate bipolar transistor) and MOSFET (metal oxide semiconductor field effect transistor). It is found in a research that

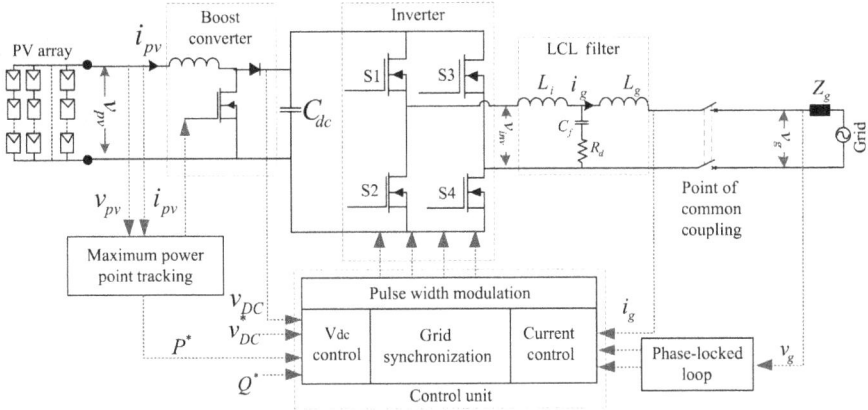

Figure 8.1   A typical grid-connected PV system

around 34 per cent of failures in power electronic systems are due to the switching devices and soldering-joint failures [6]. In addition, the DC-link capacitor is another most vulnerable component of the system [7]. Faults in these switching devices can be of two types. First type is failures that occur suddenly such as sudden overvoltage or overcurrent, or sudden temperature increase. Second type is the failures that occur by undergoing through gradual deterioration over a long period.

A typical grid-connected PV system is shown in Figure 8.1. It consists of a PV array as the generator of DC power; further, there is a DC-link capacitor at the output of PV. For applications requiring a very-high-quality and reliable power, the system requires a DC/DC converter for changing the PV output voltage level. This DC/DC converter is responsible for implementing the maximum power point technique to extract maximum power from the PV generator [8]. In addition, there is a DC/AC power electronics converter, i.e., inverter for flexible and efficient power conversion from DC to AC. The fault may occur anywhere in the system and can pose many challenges to the system. These challenges include maintaining reliability, maintaining optimum power quality in the event of a fault, dealing with the incurred energy loss, detecting and identifying the nature of fault. Therefore, for better and improved reliability, it is required that the system should have a mechanism for prior forecast of possible occurrence of fault. Sometimes the fault detection is delayed and sometimes the fault may remain undetected. Both these events may lead to catastrophic consequences. Essentially, the major challenge is continuous system operation in the event of fault. It is required that the system must have a fault tolerant mechanism in place so that the system does not shut down even in the event of an unexpected failure [9].

Failure can occur at the DC side or at the AC side. The DC side failure is caused due to the fault at the PV panel, capacitor or boost DC/DC converter. The AC side is affected by the fault at inverter or fault in the filter elements. Since PV works in intermittent environment conditions, it may experience material deformation due to ageing, which affects the PV performance badly. Human errors such as wiring mistake, careless handling and manufacturing error may be another potential cause of poor operational performance of PV, which may lead to fault events. Arc fault, local hotspots and cell material sensitivity to temperature are various other possible fault occurrences [10]. However, the most hazardous faults for the solar PV are the ground faults. Careful ground protection schemes must be employed for the PV systems. They may depend upon several factors including the plant size, installation type and the location of the plant. Irrespective of the ground protection method employed, sometimes there may be ground fault occurrence due to unavoidable reasons. These reasons can be due to damage of the cable insulation, ageing, corrosion, unintended short circuit, moisture entrance, etc. These fault events may lead to the system shutdown. Hence, there needs to be proper fault tolerance mechanism along with the ground protection strategy. Bypass diode failure, open circuit fault, line-to-line fault, double ground fault, etc. are other failure events that may happen in the PV systems [11]. Among these, line-to-line faults are quiet dangerous. Unintended low impedance path between two points on a PV panel may cause line-to-line fault. It may lead to change in the current path and may further lead to fire in worst-case scenarios [12]. Generally, fuses and overcurrent protection methods are utilized to avoid line-to-line failure events. Arc faults may also be harmful for the PV system's efficient operation. Arc faults may be categorized into series and parallel arc failure events. Loose screws, misconnection in the wiring, increased thermal stress, etc. may cause these events [13]. Therefore, the reliability of PV systems play a major role in ensuring the efficient and continuous operation of the overall system.

As discussed above, the failure analysis is of utmost importance for a robust and reliable system. Recently, fault tolerant techniques for grid-connected PV systems have become a major research hotspot. Various fault tolerant techniques have been reported in the literature [14]. They can be categorized into two types: (1) fault tolerance based on device redundancy and (2) fault tolerance based on modification in control algorithm. Once a fault is detected, the next step is the fault isolation. It is important to isolate the faulty part from the healthy part to ensure the safety of the working personnel. In addition, the next step is to apply the adopted fault tolerance algorithm. It is required that the fault tolerance algorithm should be utilized considering the overall system requirements along with the inverter configuration. Whenever a fault occurs in the system, the inverter output experience a change, i.e. the output voltage and current waveforms are distorted. To diagnose the type and nature of fault, these distorted waveforms are analysed. There are various fault diagnostic techniques proposed in the literature. Park's vector method is proposed in [15], which requires the calculation of magnitude and phase angle of current vectors along with the analysis of their space vector trajectory. The normalized and improved park's vector method is also proposed [15]. Centroid-based method [16], current pattern recognition method [17], spectrum analysis method [11], wavelet-fuzzy method [18], logic-based method [19] and model-based methods [20] have also been proposed. Recently, the artificial intelligence (AI)-based methods are increasingly adopted for fault diagnosis and identification, and they are considered to provide immensely improved results. These methods require signal processing as the first step towards fault diagnosis. Signal processing may include noise removal from signal, removing redundant signal information and normalizing signal data. Further, feature extraction is also a part of signal processing. Feature extraction is required to extract useful information. For signal processing, various techniques can be used such as discrete Fourier transform (DFT), fast Fourier transform (FFT), short-term Fourier transform (STFT) and wavelet transform (WT) [21] depending upon the system requirements. Among these, WT-based techniques are mostly preferred since they provide better signal time-frequency characteristics compared to other techniques. After the signal processing part, the next step is to apply AI-based techniques for fault detection and classification. ML and deep learning-based intelligent classification techniques provide high accuracy and efficiency as compared to other techniques [22]. Supervised, unsupervised or semi-supervised learning can be utilized according to the requirements. One of the supervised learning-based fault classification methods is artificial neural network (ANN) [23]. However, ANNs may sometimes suffer from overfitting and may fail to predict correctly. An improved version of ANN, multi-layer perceptron network (MLPN), has also been proposed to overcome the issues associated with the ANN [23]. Decision tree [24] and random forest [25]-based methods have also been proposed in literature. Other ML-based techniques such as *k*-nearest neighbour (K-NN) [26] and support vector machine (SVM) [27] have been presented for grid-connected PV systems. Recently, deep learning-based fault classification techniques, which utilize convolutional neural network (CNN) [28], have taken precedence over other techniques since they provide outstanding accuracy. A grid-connected PV system with fault diagnostic approach is shown in Figure 8.2.

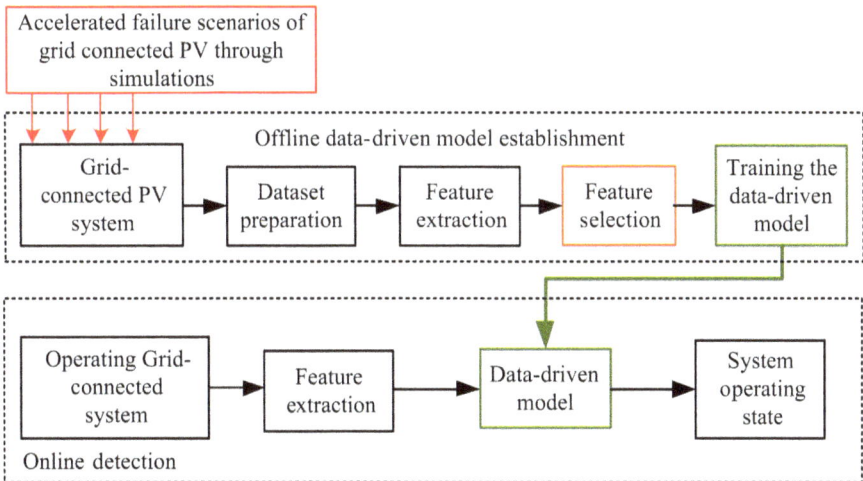

Figure 8.2   Fault diagnostic block diagram for grid-connected PV system

8.2   Components of power electronic converters

Today, PV is considered a promising energy source that is utilized not only in power sector, but in transportation, lighting, construction and residential sectors as well. It is pollution free, green and abundantly available. However, it provides DC power and requires an inverter for DC/AC power conversion. With inverter as an interface, PV can be connected to grid and other AC loads. The inverter is responsible for providing clean sinusoidal output with the desired voltage and frequency. Inverter must be able to synchronize with the grid for a proper and efficient grid connection. Sometimes, the system might require a boost DC/DC converter for enhancing the output of PV array. Both these power electronic converters, i.e. DC/DC converter and DC/AC converter (inverter) consist of semiconductor-based power switching devices as a major component. With the advancement in semiconductor technology, IGBT (insulated gate bipolar transistor) have become the most suitable switching devices for power electronic converters [29]. The IGBT has combined the advantages of both BJT (bipolar junction transistor) and MOSFET. It has high switching speed, high input impedance and low saturation voltage. However, along with the DC-link capacitor, it is the most vulnerable component of the grid-connected PV system. The majority of failures in the power electronic converters occur due to the IGBTs and DC-link capacitors. Therefore, the study of IGBT's electrical and thermal characteristics is of paramount importance.

IGBTs have incorporated huge developments in the last few years in terms of dimension, cost, thermal efficiency, reliability, etc. The IGBTs now come with higher power densities owing to the innovative assembly and packaging techniques. In addition, the power outputs of commercial IGBTs can be as high as millions of watts. It is expected that in future also, the IGBT power output will further increase. Today, not only IGBTs have undergone structural and operational advancements, they also have better and improved cooling capabilities, which make them a great choice for power electronics converters. IGBTs are widely used in many critical applications such as switched-mode power supplies (SMPS), uninterrupted power supply (UPS) and inverters. These applications require efficient modulation techniques for the proper operation of the switching devices. However, despite the continuous improvement in the IGBT characteristics, the reliability concerns still exist [30]. Due to the electrical, mechanical and thermal overstress, the IGBTs are susceptible to failure. IGBT failures may be of two kinds – one is the wear out failure and other is catastrophic failure. The wear out failure may occur due to the bond-wire lift-off, bond-wire heel crack, solder-joint failure, avalanche breakdown, latch-up, secondary breakdown, etc. These wear out failures are experienced owing to the long-time operation of the device. The other reason might be the power cycling [31]. The catastrophic failure in IGBTs occur due to the overstress events such as sudden increase in voltage, current or temperature. When the device capacity is not enough to handle this sudden change, the device fails.

8.2.1   IGBT failure

The requirement of modern power electronic systems is to meet the specific reliability and robustness level. To fulfil these requirements, it is desired that the efficient

Table 8.1   IGBT failures and their causes

Failure	Failure categories	Causes of failure
Catastrophic failure	High-voltage breakdown	High-voltage spikes
	Latch-up failure	High dv/dt during turn off
	Electrostatic discharge (ESD)	High-voltage application to the gate
	Secondary breakdown	High-current stresses
	Mechanical overstress	Surface metallization
Wear out failure	Bond-wire lift-off	Mismatch between CTEs of aluminium (Al) and silicon (Si)
	Bond-wire heel crack	Ageing
	Solder-joint failure	Mismatch between CTEs of Copper and Si
	Gate-oxide degradation	High temperature and high electric field

cooling and proper thermal efficiency be ensured for the IGBTs. From the reliability perspective, the junction temperature of IGBT play a vital role. It is also important to analyse the heat losses in the IGBTs for any power electronics converters. The heat losses can be categorized into two categories – conduction and switching losses. The conduction losses depend upon the conducting current when there is on state voltage drop across the IGBT. The switching losses occur due to the turning on and off stages of the device depending upon the duty cycle, switching frequency, current and the voltage of the device [32]. The mechanical and thermal losses might also be there which are caused due to the overstress events. The various kinds of failures occurring in the IGBTs are summarized in Table 8.1 and detailed here.

8.2.1.1   Wear out failure

For ensuring the reliability of the power electronic converters, the study of wear out failure in the IGBTs is very important. The IGBT material experiences these failures due to thermo-mechanical fatigue stress. These failures can further be categorized as:

Bond-wire lift-off – The bond-wire lift-off failure occur due to mechanical events. The bond-wire can be raptured due to the mismatch between the coefficients of thermal expansion (CTE) of Al and silicon (Si). Due to the temperature swings and difference between the CTEs, the bond-wire crack occurs and due to the growth of the bond-wire cracks, bond failure occurs. The strain difference between the two materials causes stress and that stress depends on the temperature. This results in the initiation of crack and is further propagated by the stress-strain energy loss [33]. This hysteresis energy loss is represented by the temperature swings. When the junction temperature swings are sufficiently high, the crack travels from the sides of Al wire to the centre. That is where the bond-wire lifts off.

Bond-wire heel crack – The bond-wire heel crack is a rare event in the advanced IGBT. After long time operation of the IGBT, bond-wire heel crack can be observed and is considered as the fundamental reason behind the failure mechanism for power electronics converters [34]. It can be detected through the measurement of saturated collector to emitter voltage (V_{CEsat}). The criteria to detect this particular failure is an approximate increase of 5 % in V_{CEsat}.

Solder-joint failure – Solder-joint failure occurs very frequently and is one of the major causes of the wear out failure of IGBT. Copper has a higher CTE than Si. This difference in CTE results in a shear stress between the Si die and copper substrate, which further leads to the formation of cracks or voids. These voids result in the effective area of heat dissipation being decreased leading to the concentration of heat on the surface [35]. This further accelerates the formation of voids leading to extremely localized heating due to increased thermal resistance. This together with the power cycling process is responsible for solder fatigue. The solder-joint failure adversely affects the collector to emitter ON voltage (V_{CEon}) and its value reduces in the linear region of operation of the IGBT.

Gate-oxide degradation – The gate oxide in the IGBT can be degraded over time owing to high temperatures and high electric fields. This is caused by the dielectric breakdown dependence over time [36]. Due to impact ionization, hot electrons damage the gate-oxide layer. Gate-oxide damage leads to the increase in the gate leakage current. It also adversely affects the collector to emitter threshold voltage V_{CEth}.

8.2.1.2 Catastrophic failure

Other than the wear out failures in the IGBT, there are catastrophic failures, which occur due to the excessive electrical, mechanical and thermal stress events. They adversely affect the operation of IGBT and sometimes may permanently damage the device. These failures can be further categorized as:

High-voltage breakdown – Due to high-voltage conditions, heating effects can be significant which further affects the IGBT operation. This becomes a major concern in power electronic converters. High-voltage spikes may occur due to the high decay rate of collector current. Repetitive occurrences of these spikes may destroy the IGBT during the turn-off conditions [36]. This can further lead to increase in the leakage current and high local temperature. Higher values of collector to emitter voltage (V_{CE}) or gate to emitter voltage (V_{GE}) during the turn off conditions may lead to short circuit. It then becomes necessary to ensure the suitable safe operating area (SOA) for the particular power device to meet the desired heat-sink specifications.

Latch-up failure – Latch-up occurs when gate is no longer able to control the collector current. High dv/dt during the turn-off conditions may lead to the triggering of the parasitic transistor of the IGBT further leading to latch-up. High collector current may also lead to latch-up through the turn-on of the parasitic transistor [37]. The loss of gate control due to latch-up damages the IGBT. However, there has been an improvement in the latch-up immunity of the recent devices; it needs further attention to ensure the reliability and robustness of the power electronics converters.

Electrostatic discharge – ESD is very similar to the high-voltage breakdown failure. It can cause partial puncture of the gate oxide, which allows the device to work properly for a particular operation period and ultimately leads to the failure of the device after a certain duration of time [37]. When excessive voltage is applied to the gate, it leads to the short at the gate, which further affects the operation of the device.

Secondary breakdown – The high stresses of current may lead to the local thermal breakdown of the device during the switching on and off of the device. Secondary breakdown is a kind of thermal breakdown. As the current increases, so does the space charge density at the collector-base junction, along with the decrease in breakdown voltage, in turn causing a rise in the current density [37]. This cycle continues until the time the area of the high current density region is reduced to the minimum area of a stable current filament. Consequently, the temperature of the filament increases rapidly because of self-heating leading to a sudden decrease in voltage across the IGBT.

Mechanical overstress – Due to the mechanical overstress events, atomic migration may occur which in long run may affect the reliability of the device. Surface metallization has been observed in IGBT devices, which may happen due to the Al bond pad reconstruction [37]. This can further cause unequal thermal resistance distribution, which leads to the high temperature in the device. This may even cause device local burnout if proper mitigation steps are not taken.

8.2.2 Thermal modelling of IGBT

With the continuous advancement in the switching frequency and power density of the IGBT devices, the electrothermal and thermo-mechanical analysis becomes imperative. These multidimensional studies are required to attain the optimized device operation. Sometimes, one kind of fault may induce another kind of failure in the device, which might be depending upon some different electrical traits of the device. Hence, there is a need to devise optimum holistic thermal or electrothermal or thermo-mechanical model, which is able to include overstress mechanical events, high junction-temperature events, bond-wire failures, high dielectric events, heat-sink issues and all the potential failure causes. The most widely used models are thermal models, which can easily incorporate device characteristics through the resistance-capacitor (RC) network model. These thermal models are computationally efficient and require minimal effort in extracting the junction temperature information. They can be easily simulated electrically and provide satisfactory output. The modelling should be done such that the accuracy of the model is high, and it is able to analyse the reliability parameters as well. These thermal models can be categorized here.

8.2.2.1 Analytic models

Initially mathematical thermal models were developed for the reliability analysis. However, there has been a continuous development in these models over the years and now they have become highly efficient. Based on the two-port theory, they have devised computationally efficient tools for effective junction temperature rise calculation and heat flux distributions. Boundary element method (BEM) was used for simulating an IGBT of high power with power cycling [38]. It was able to significantly reduce the complexity in the device geometry and was able to solve the heat losses calculation. The 3D finite difference methods (FDMs) was also proposed to solve the equation of heat-conduction for the thermal model of IGBT. The mechanism of thermal interface material (TIM) was also introduced for optimizing the model's geometry-based operating conditions and analysing heat dissipation of the heat sink.

8.2.2.2   Numeric models

Finite element method (FEM) based on the numeric simulation was proposed to obtain the temperature distribution of the device with a significantly high accuracy. It was based on the material properties along with the detailed geometric structure of the device. Over the years, they have incurred huge developments and have become more advanced than before. They can calculate the power loss occurring in the device along with the device temperature using coupled approaches. They can correctly model the thermal characteristics along with the model parameter sensitivity. They can further help in identifying the heat-sink characteristics to provide an efficient cooling mechanism for the device.

8.2.2.3   Network models

These are widely utilized for the dynamic and steady state thermal analysis of the power device [39]. It can be represented with the help of basic thermal resistance (R_{th}) and capacitance (C_{th}). Initially, a charge-controlled model was developed which was analytical in 1D and was able to implement the circuit simulator for the IGBT. The calculation methods were developed for heat diffusion equations and temperature distributions. This method was able to easily to connect to the electrical network, which further helped in fast dynamic calculations along with the estimation of the electro-thermal characteristics. During the switching cycle of the device, this method was able to accurately determine the IGBT temperature rise. Two networks classified under this particular model are Foster and Cauer RC networks [40]. These networks have some structural differences but accuracy-wise they reflect almost the same behaviour. These models have very high accuracy with low time-consumption, and they can deal with thermal, electrical as well as physical properties of the power devices.

8.2.3   Cooling measures

The reliability of the switching devices such as IGBT can be enhanced if a proper cooling strategy is adopted. One of the major design considerations for IGBTs is the heat-sink considerations since it directly affects the thermal management and hence, reliable operation of the IGBT. Many failures of IGBT devices can be prevented if there are improved cooling measures are taken. The heat-sink operation is not influenced by the geometrical structure; rather its operation is mainly affected by the material used for cooling. Air cooling methods are mostly utilized since they are low cost and easily available. However, they have poor cooling efficiency as compared to the liquid cooling methods. There are direct liquid-based cooling solutions proposed in literature such as heat-sink cooling method and spray cooling method. These cooling methods perform much better than the air-cooling methods since they have the ability to reduce the thermal resistance by 30 per cent [41]. The conventional indirect cooling was based on heat sink having power module mounted on it through thermal grease leading to high thermal resistance. Nevertheless, with the direct liquid cooling method, there is no thermal grease and hence, reduced thermal resistance and better cooling efficiency. There are micro-channel cooling solutions proposed in the literature. They require minimum amount of coolant, have compact size and possess unique cooling behaviour. Two-phase forced convection cooling solution is also proposed, and it provides better heat transfer coefficient compared to other methods. These solutions are preferred where there are high reliability requirements since they have superior characteristics of maintaining uniform device surface temperature. Jet impingement and spray cooling methods are largely employed for high-power applications. They have amazing cooling properties and are one of the best solutions for larger surface areas since they can easily eliminate large amount of heat. One of the modern and novel cooling methods is hybrid solid and liquid cooling solution for isothermalization of switching devices [42].

8.2.4   DC-link capacitor failure

Capacitors are one of the most unreliable components of any electrical system after the semiconductor-based switching devices. For power electronics converters applications, electrolytic capacitors are popularly used since they are cost effective. However, they suffer from some major issues such as low filter-bandwidth, sensitivity to variations in temperature and short life. The MPPF (metallized polypropylene film) capacitors are a better choice for power electronics applications [43]. They have goof filter bandwidth, low ESR (equivalent series resistance) and self-healing property. Along with these advantages, they are rarely affected by the surrounding temperature variations. Nonetheless, both these capacitors are vulnerable to fault occurrences.

There can be various kinds of ways in which capacitors fail. Early failures may occur due to the issues in manufacturing process. Random failure events are very rare in case of capacitors. Mostly, the capacitor failures are wear out failures. Due to longer exposure to temperature variations, partial discharges and ageing may cause capacitor failures. They can also fail due to electrical and mechanical overstress events along with general environment wear and tear events. The gradual breakdown of dielectric may also lead to fault current and reduction in capacitance, and ultimately complete failure. Capacitors may also fail due to improper operation, e.g. continued overvoltage or overcurrent events. Open-circuit fault occurrence may not cause the failure to spread further, but short circuit fault events are regarded as more harmful for the capacitor health. Hence, it is important to take some action in the initial phase only, otherwise short-circuit fault may further propagate and may lead to failure of the neighbouring devices. For capacitors mounted on PCB (printed circuit board), corrosion of leads may occur due to the external materials leading to failure. Repeated overstress events are very harmful for capacitors as they may lead to increase in ESR value causing more heat losses.

Electrolytic capacitors are one of the weakest components of the grid-connected PV systems. Their equivalent circuit consists of a capacitor in series with an equivalent resistor and inductor. These capacitors are susceptible to fault events in the system and may be affected by the external temperature changes. The primary wear out failure may be caused by the vaporization of the electrolyte between the cathode and anode terminals of the capacitors. This vaporization leads to effective decrease in the electrolyte area causing decrease in capacitor value and increase in the ESR value. This leads to rise in temperature and further vaporization of the electrolyte causing the repetition of the same process. There are standards, which suggest that for the electrolytic capacitor to fail completely, the ESR increment should lead to the doubling of ESR value and the capacitor value should decrease by 20 per cent. Various failure detection techniques have been proposed in the literature to identify and diagnose the failure mechanism in the capacitors. These techniques are based on identifying the corresponding ESR and capacitor value change and the temperature variations.

8.2.5   Power diode failure

Si power diodes serve as one of the important components of the power electronic converters. However, they may suffer from failures under extreme stresses of voltages or currents. It is observed that the major failures occur in power diodes due to high-voltage spikes. Even if the voltage or current spike does not have enough energy to produce temperature rise of the device, it can still cause damage to the device [44]. This kind of failure can be detected through the analysis of current density since it will be higher at a certain point. Other major diode failure occurs due to secondary breakdown. The power diode VI characteristics consist of positive resistance in the low-current region and high-current region has negative resistance characteristics. The current density in both the regions remain balanced. If due to any reason, the current density imbalance is observed, then a higher current density appears at a certain concentrated area, which can give rise to secondary breakdown. This secondary breakdown eventually leads to the failure of diodes. Therefore, for the proper operation of the grid-connected PV systems, it is imperative to consider the failure mechanism of the power diodes.

8.3   Failure mechanisms of power semiconductors

8.3.1   Failure mode, mechanisms and effects analysis

Failure mode effect analysis (FMEA) is a method for developing comprehensive lists of failure modes for a system and analysing the effects of the failure mode to the larger system. Failure mode effect and criticality analysis (FMECA) expands upon FMEA by introducing a criticality metric through which the failure modes are ranked based on the severity, occurrence and detectability of each failure mode. The criticality analysis allows engineers to focus on these critical failure modes, identified by a risk priority number, to reduce the effects to the end user or system manufacturer.

Further, these methods for identifying failure modes and were first established in the 1940s by the United States military. The US Department of Defense published and updated MIL-STD-1629A – Procedures for Performing Failure Mode, Effect and Criticality analysis [45]. The Society of Automotive Engineers (SAE) published ARP926 in the 1960s [46]. The electronics industry had used it but formally codified adopted the FMEA process when Joint Electron Device Engineering Council (JEDEC) Solid State Technology Association published of JEP131 [47]. The JEP131 document defines FMEA as: 'an anticipatory thought process designed to utilize as much knowledge and experience of an organization as possible toward the end of addressing potential issues defined in a new project. The objective is to reduce the probability that a customer is exposed to a potential product and or process problem by performing a thorough risk analysis' [48]. FMEA should be completed by a group of subject matter experts for the systems identified. In this work, the JEP document is used as a primary source since it is particularly developed for semiconductor components.

The development of an FMEA is as follows. First the system which is to be analysed must be clearly defined. Then the system should be broken down into subsystems either in a functional, geographical/architectural, or combination of the two. In this step, all the functions of the subsystems should be identified. Next, identify all the possible failure modes that the subsystem can experience which may be done using a variety of techniques including testing, engineering judgement and simulation. Then, for each mode, identify the possible causes of failure for each of the failure modes. Here, a life cycle profile for the product should be developed to help understand the various stresses the component may see not only during operation but also during manufacturing, storage and transportation. Finally, identify how the failure affects the end user. Like the life cycle profile, this information is application-specific. Once these steps have been taken the FMEA is completed.

A final step for FMEA and failure mode mechanisms (FMM) and effects analysis (FMMEA) is to attempt to prioritize the failure modes, mechanisms and effects to allow for effective usage of resources to address reliability concerns. Here again, the analysis is highly application specific. As opposed to FMEA which identifies the high-risk-failure modes to update the design and reduce risks to acceptable levels, FMMEA takes the FMEA an additional step and identifies the failure mechanisms associated with failure causes and modes [49]. For failure mechanisms, relevant failure model(s) can be identified which can illustrate how the stress leads to the failure of a system. Failure mechanisms are highly dependent on the materials, geometries and stresses within a system.

In the literature, Patil *et al.* [50] have published a FMMEA for Si power devices, see Table 8.2; however, this FMMEA is limited to only discrete IGBT parts, power cycled at high mean temperatures with large junction temperature swings [50]. The FMMEA was developed for the purposes of identifying failure precursor parameters for prognostics applications and it served that function well. Such an FMMEA is limited in several respects, first as will be discussed in failure mechanism criticality analysis, FMMEA

Table 8.2   Potential failure modes, causes and mechanisms in literature

Potential failure modes	Potential failure causes	Potential failure mechanisms
Short circuit, loss of gate control, increased leakage current (oxide)	High temperature, high electric field, overvoltage	Time dependent dielectric breakdown (V_{th}, g_{th})
Loss of gate control, device burnout (Si die)	High electric field, overvoltage, ionizing radiation	Latch-up $(V_{CE(ON)})$
High leakage currents (oxide, oxide/substrate interface)	Overvoltage, high current densities	Hot electrons (V_{th}, g_m)
Open circuit (bond-wire)	High temperature, high current densities	Bond-wire cracking, lift-off $V_{CE(ON)}$
Open circuit (die attach)	High temperature, high current densities	Voiding, delamination of die attach $(V_{CE(ON)})$

requires application knowledge, therefore this is not truly an FMMEA and it makes no effort to account for failure mechanism criticality. Second, Patil *et al.*'s list does not include all relevant failure mechanisms for Si power devices in the reasonably expected operating conditions.

8.3.2   Power semiconductor failure mechanisms

This section provides a description of relevant failure mechanisms that have been observed in Si power devices and a table summarizing the failure modes, sites, causes and mechanisms. Once a life cycle profile has been identified, a list of failure modes and mechanisms that may be precipitated from the stresses present in the life cycle profile must be established in an FMMEA. While there are many potential sources of thermal, mechanical and electrical stresses on the power semiconductor components, the mechanisms which may be precipitated are dependent on the physical characteristics of the power semiconductor component. This section provides an overview of the failure mechanisms which have been reported in power semiconductors. It is not intended to be a comprehensive listing of all the relevant literature on each possible failure mechanism; however, a basic description and several sources are identified for each of the mechanisms.

8.3.2.1   Aluminium reconstruction

The die metallization for power semiconductors is typically Al. Additionally, if the substrate is direct bonded aluminum (DBA) then there is Al metallized on the ceramic substrate. Due to thermal cycling of the component due to joule heating, switching losses, changes in the temperature of the environment and the mismatch of thermal coefficients of expansion between the Al and Si and ceramic substrate and thermo-mechanical stresses are generated within the package. These stresses can be significant enough to cause yielding of the Al metallization, causing it to buckle and form hillocks. This mechanism is referred to as Al reconstruction [51–53]. The reconstruction of Al can increase in the resistance of the metallization layer. Al reconstruction can be exacerbated by electromigration which can happen if significant current densities are present in the metallization. Al metallization that is coated with a passivation layer, typically silicon nitride, has been shown to resist reconstruction; however, bond pads, which make up a significant portion of the total metallized area, are not passivated and therefore remain unprotected.

8.3.2.2   Bond fatigue

Bond-wire fatigue is also a thermo-mechanically driven failure mechanism. Due to the large diameter of wire used to handle high current densities in power devices, Al bond-wire must be wedge-bonded on both ends of the connection unlike gold or copper bond-wire which are typically used in low-power applications with smaller wires that can be ball bonded. Fatigue of these wedge bonds occurs due to joule heating, switching losses, changes in the temperature of the environment and the mismatch of thermal coefficients of expansion of the Al wire and the Si die [51–53]. Bond-wire fatigue manifests itself as either lift-off or heel cracking. The heel of the wedge bond acts to as a stress concentrator and a crack propagates through the heel. In bond-wire lift-off, the wedge delaminates from the surface of the bond pad at the heel and toe sides of the wedge bond and propagates inward. Bond-wire fatigue results in an increase in the on-state resistance of the power device and can lead to an open circuit.

8.3.2.3   Die-attach fatigue and delamination

Another location of thermo-mechanical stress due to CTE mismatch is the die attach which connects the die to the substrate. As the die is vertically conductive, the die attach must be conductive as it is part of the electrical path of the power semiconductor component. Similar to the Al metallization and wire bonds, the die attach is in intimate contact with the die and undergoes thermo-mechanical fatigue and possible delamination [54, 55]. Delamination occurs when the separation is between the die itself and the die-attach material; however, fatigue can occur and propagate through the die attach. Additionally, the delamination of the die attach increases the thermal resistance of the die attach, decreasing the ability of the die attach to dissipate heat generated at the die. Increased thermal resistance results in higher die temperature impacting the electrical characteristics of the device.

8.3.2.4   Substrate cracking

The substrate itself is a possible location of failure within a power module. The ceramics in the direct bonded copper (DBC) and DBA substrates can crack when subjected to thermo-mechanical cycling due to operational and environmental loading [56]. The substrate acts to insulate the conductive paths of the power package from the heat-sink or other cooling mechanisms. Therefore, when cracking occurs, the insulation properties of the ceramic break down and a reduced insulation strength is observed. Depending on the electrical connection of the heat sink, this can create a significant leakage path within the power package.

8.3.2.5   Bond-wire melting

While the die is the most significant source of joule heating within the power package, the parasitic resistance of the bond-wires also acts as a heat source. The bond-wires are encapsulated in either silicone gel or an epoxy moulding compound (EMC) depending on whether or not the package is a discrete or a module. Both silicone gel and EMC are poor conductors of heat. This means heat generated within the bond-wires is difficult to dissipate and the wire can see elevated temperatures. If the power package is used near or at its current rating and effective measures are not taken to cool the package, the bond-wires can melt [57]. The result of this failure is an open circuit of the transistor.

8.3.2.6   Die-attach voiding

Another failure mechanism related to the die attach is voiding of the die attach [58–60]. Small amounts of voiding are residual in the die attach from the manufacturing process. Power cycles grow and coalesce the smaller distributed voids which are initially present in the solder. Like die-attach fatigue, die-attach voiding results in an increase in the on-state resistance of the power package and increases the thermal resistance of the packaging, thus increasing the junction temperature of the package.

8.3.2.7   Aluminium corrosion

When moisture is present within the package, corrosion of the Al wire bonds and bond pads can be of concern [61, 62]. In the presence of moisture, Al reacts to form $Al(OH)_3$ which passivates its surface and passivates the Al. This passivation layer can become soluble in the presence of contaminants such as halogens. During power cycling of a component, the encapsulant layer can delaminate from the base plate thus developing a path for moisture and contaminants to ingress into a component. Similar to the thermo-mechanical fatigue mechanisms, corrosion would cause an increase in the on-state resistance of the power semiconductor and is a wear out failure mechanism.

8.3.2.8   Latch-up

There are parasitic circuit elements associated with the power package as semiconductors and metals have non-ideal material properties associated with them. One particularly harmful parasitic element within power components are parasitic thyristors. Thyristors are switches which only require a voltage or current pulse to turn on, unlike other switches such as MOSFETs or IGBTs which require a voltage on a gate be maintained for the switch to be in the on-state. Thyristors continue conducting after the removal of the gate voltage or current pulse until the electric potential between the anode and cathode is zero. Power semiconductor device have a parasitic thyristor incidentally built in.

Switches such as MOSFETs and IGBTs are expected to conduct only when a gate voltage is applied and, therefore, the activation of the thyristor may cause the loss of gate control of the device. Such an event is referred to as latch-up and is observed as a short circuit between the conduction terminals [63]. The thyristor is activated when the current exceeds the so-called latching current of the device. In an IGBT, this overcurrent forces current into the base of the parasitic NPN transistor, as the local high-current density in the P region at the base increases the resistance locally.

Latch-up of a device does not inherently cause a destructive failure of the component. In the unlikely event that the latch-up event is detected and measures are taken to remove the current from the device, the event can be stopped and the device will function normally. However, if the latch-up event is not identified, it can lead to a thermal runaway causing the device to burn out. Latch-up is an overstress failure mechanism and is observed as a short circuit of the device.

8.3.2.9 Avalanche breakdown

The avalanche breakdown [64, 65] mechanism can precipitate, often during switching, when the drain-source or collector-emitter voltage exceeds the breakdown voltage of the power device. Electrons within the device gain sufficient energy to impact atoms within the device and ionize the atoms and releasing additional electrons. If these impacts continue, the device can 'avalanche' as an increasing number of electrons are freed and able to impact atoms to free addition electrons. Avalanche breakdown often occurs during switching of a device when the inductance of the power semiconductor or the system within which it is operating, creates a voltage spike on the system. Avalanche breakdown manifests itself as a short circuit of the device and is considered an overstress mechanism.

8.3.2.10 Partial discharge

The silicone gel that encapsulates the metallization's and bond-wires within power modules is used to increase the breakdown strength of these conductors. However, due to the high-voltage and geometries of the conductors the electric field within the package can still be enhanced and cause a partial discharge within the silicone gel [66, 67]. Over time, these discharge events can develop a carbonized, conductive path within the gel leading to increased leakage. Locally, partial discharge can cause bubbles in the gel to form due to the local heating events. Partial discharge within the gel is observed as increased leakage, developing towards a short circuit and is a wear out mechanism.

8.3.2.11 Electrochemical and silver migration

Often die attaches contain silver, as a sintered silver paste or a tin–silver alloy solder, which has a strong propensity to migrate under a variety of conditions [68, 69]. In the presence of moisture, silver and other metals show some slight solubility. If an electric field is also present, the silver and other metals will migrate from the anode to the cathode through electrochemical migration. At the cathode, the silver will deposit and form dendritic structures back towards the anode. Given time, these dendrites can grow long enough to short the cathode and the anode. Mass transport of silver can also happen through corrosion, particularly in the presence of sulphur. Unlike electrochemical migration, in silver migration due to corrosion, there is no expectation of growth direction. Silver migration and electrochemical migration manifest themselves in an increased leakage current and are wear out failure mechanisms.

8.3.2.12   Dielectric breakdown

It is essential that the insulated dielectric that forms the gate of voltage-controlled devices maintains dielectric integrity for the device to operate. If the electric field through the gate exceeds the dielectric strength of the insulating material, then the terminals will short and permanent damage will be done to the gate [70]. For Si devices, the gate is made of silicon dioxide. Dielectric breakdown can occur due to an overvoltage event on the die for a short period of time. One possible cause of this overvoltage is an ESD. Such cases would be overstress events and likely result in a shorting of the gate to one of the conduction terminals.

8.3.2.13   Time-dependent dielectric breakdown

Dielectric breakdown can also occur over time through a process called time-dependent dielectric breakdown [71]. One leading explanation for this mechanism is that Si–Si bonds within the dielectric are weak and over time the application of an electric field breaks down these bonds, creating locations within the dielectric through which electrons can jump to and travel through the insulating gate. Time-dependent dielectric breakdown manifests itself as high gate leakage current and is a wear out mechanism.

8.3.2.14   Hot carrier injection

Some electrons may gain sufficient energy while travelling through the MOS channel to be able to tunnel through the gate-oxide layer [72]. These electrons become 'hot', referring to their individual speed, and as a consequence energy, as opposed to the bulk temperature of the device itself, as they travel along through the gate channel when the device is conducting. These electrons can cause impact ionization near the end of the channel which can produce electrons which can inject themselves into the gate dielectric. All these steps can cause damage at the interface of the Si and silicon dioxide or allow the carriers to become trapped within the dielectric itself. Hot carrier injection (HCI) causes parameters associated with the gate such as the gate threshold voltage to shift. Under HCI damage, gate threshold voltage would drift higher, requiring higher gate voltage to be applied to achieve the same level of conduction in an otherwise healthy device. HCI is reported to be more common at low temperatures, unlike most other mechanisms which are thermally accelerated. At lower temperatures, lattice scattering is reduced allowing longer free paths for electrons to accelerate, gaining energy to create hot carriers. HCI is a wear out mechanism.

8.3.2.15   Competing failure mechanisms

The failure mechanisms discussed in this section are not all independent of each other. In many cases one failure mechanism may lead the device to failure through another mechanism. For example, the degradation of the die attach can cause a latch-up event on the die. Power cycling through delamination and voiding causes both an increase in electrical and thermal resistance. This leads to an increase in the temperature of the die during operation as more power is dissipated due to the resistance increase and heat cannot leave the package as easily. Additionally, as portions of the die attach have 'disconnected' from the die, current crowding occurs, leading portions of the device susceptible to latch-up failure event though the device itself is still conducting the same amount of current. In such an instance, it is evident that one failure mechanism drove the device to failure through another mechanism. Due to the potential of failure mechanisms to convolute each other, it is important for engineers to understand such mutually accelerating factors when designing systems.

8.3.3   Power semiconductor failure modes and mechanisms

In Section 8.3.2, a list of relevant failure mechanisms was developed for Si power semiconductors. Such a list that we name FMM is the foundation of any FMMEA which may be developed for a component. This list, as well as information from the life cycle profile in the context of the application should be combined to establish an FMMEA. The failure mechanisms discussed in Section 8.3.2 are listed in Table 8.3.

Table 8.3 is organized to include the failure mode, location, causes and mechanism. Systems integrators will find this information useful for identifying the failure causes and mechanisms that should be considered in the design of the system. For example, if they are aware that moisture will be present in the application, they should consider relevant measures to prevent ECM and silver migration. The corollary to this is that if they are confident that no significant moisture will be present then such measure may not be necessary in the design of the system.

Another possible application of this table is for failure analysis engineers. Based on the information that they establish during the failure analysis; this table can help lead the failure analysis team to identify the cause and mechanism associated with the failure. Once the cause has been identified, the proper steps can be taken to reduce the likelihood of future failures or identify risks for fielded systems.

8.4   Data preparation and feature extraction

The main objective of the proposed fault classification methodology is to determine the suitable wavelet coefficients to detect and classify the PV system faults accurately.

8.4.1   Wavelet transform

Conventionally, FT has been widely used as the main signal analysis technique that can provide an amplitude-frequency domain representation. However, the main

Table 8.3   *Failure modes and mechanisms of Si power devices*

Potential failure mode	Potential failure location	Potential failure causes	Potential failure mechanisms
Short circuit	Collector-emitter path (die)	Collector-emitter current above latching trigger current, high temperature, cosmic rays	Latch-up (overstress)
		Collector-emitter voltage exceeds breakdown voltage, high-frequency switching, unclamped inductive switching	Avalanche breakdown and secondary breakdown (overstress)
	Gate oxide (die)	Gate voltage exceeds the breakdown voltage of the gate	Electrical overstress and ESD (overstress)
	Encapsulant (package)	The electric field between bond-wires exceeds the dielectric strength of encapsulant	Partial discharge (overstress)
Increased collector-emitter leakage current	The periphery of die (package)	Presence of moisture, high temperature, mobile ions, high electric field	Electrochemical migration (wear out)
		Presence of silver within package, moisture, high temperature, high electric field	Silver migration (wear out)
Reduction of dielectric strength	Insulating substrate (package)	Temperature and power cycling, CTE mismatch	Substrate cracking (wear out)
Increased gate leakage current and gate threshold voltage	Gate oxide (die)	Prolonged gate voltage application, high temperature	Time-dependent dielectric breakdown (wear out)
		High MOS-channel currents, low temperature	Hot carrier injection (wear out)
On-state resistance increase (may develop into open circuit)	Bond-wire (package)	Temperature and power cycling, CTE mismatch	Bond-wire cracking and lift-off (wear out)
		Presence of moisture and contaminants such as halogens	Al corrosion (wear out)
	Surface metallization (die and package)	Temperature and power cycling, CTE mismatch	Al reconstruction (wear out)
Open circuit	Die attach (package)	Temperature and power cycling, CTE mismatch	Voiding, delamination of die (wear out)
	Bond-wire (package)	High temperature due to power dissipation	Bond-wire melting (overstress)

limitation with the FT is the lack of time-domain related information. To overcome this limitation, the STFT (8.1) and WT (8.2) have emerged as time frequency transforms which can provide both time and frequency information of the analysis signal.

$$STF[e,g] = X[k] = \sum_{n=-\infty}^{\infty} x[n] g[n-e] e^{-j\omega n} \tag{8.1}$$

where x is a time domain signal, X is the transformed signal in the frequency domain, g is the window function, e is the index to define the size of the fixed window function and ω is the angular frequency.

$$X[\varphi,v] = \frac{1}{\sqrt{v}} \sum_{t=-\infty}^{\infty} x[t] w\left[\frac{t-\varphi}{v}\right] \tag{8.2}$$

where $x[t]$ is the target signal, $w[t]$ is the chosen wavelet and v and φ are the scale and shift parameters, respectively.

Generally, the WT represents a signal-processing tool that was applied to capture the features contained in the signal. There are many methods for applying WT, namely, Hilbert–Huang transform [73], Wigner-Ville distribution [74], stationary WT [75], continuous WT [76] and discrete WT (DWT) [77]. It is considered that DWT offers an exact representation for any given signal, providing frequency sub-bands at different resolutions. This phenomenon provides a great advantage for discrete wavelets over the continuous WT. It also provides significant reduction in the computational complexity by just computing the wavelet coefficients at the frequency sub-bands. Therefore, the computational complexity for the discrete wavelets is only $O(n)$, is significantly less compared to the continuous wavelets, with n as data size. The major advantage of DWT is its extensive library of the wavelet functions, making it adaptable for transient analysis. This provides different resolutions for time-frequency spectrum. Table 8.4 depicts the computational time for various WT techniques.

For every level of wavelet decomposition j, the discrete-time signal is decomposed into approximation wavelet coefficients Ca_j and detailed wavelet coefficients Cd_j. The approximation and the detail wavelet coefficients can be computed using the following:

$$Ca_j(e) = \sum_{l_e} f_{l_0}(l_e - 2e) Ca_{j-1}(l_e) \tag{8.3}$$

$$Cd_j(e) = \sum_{l_e} f_{l_1}(l_e - 2e) Ca_{j-1}(l_e) \tag{8.4}$$

Table 8.4   Computational time of various wavelet transform analysis techniques

Technique	SWT	DWT	HHT	CWT	WVD
Computational time (s)	0.195	0.004	0.241	0.245	0.08

where f_{l_0} is the low-pass filter and f_{l_1} is the high-pass filter. The energy of the wavelet coefficients at the details g_{d_j} and the approximations g_{a_j} of the j_{th} decomposition level can be calculated using the following:

$$g_{aj} = \sum e\, |Ca_j\, (e)|^2 \tag{8.5}$$

$$g_{dj} = \sum e\, |Cd_j\, (e)|^2 \tag{8.6}$$

8.4.2   Harmony search algorithm

The harmony search algorithm (HSA) is a meta-heuristic optimization technique, which was developed by Geem *et al.* [78], and has been recently utilized in several power system benchmark with success [79]. The HSA is a simple concept and is easy to implement since it requires less parameters with simple mathematical analysis. Furthermore, the need for setting an initial value of decision variables is completely eliminated. The limitations of conventional techniques, mainly the need for initial values and considerable gradient details, were highlighted in [80]. Furthermore, the study in [80] concluded that the HSA was capable of providing a superior accuracy in comparison with present meta-heuristic optimization algorithms.

The optimization steps of the HSA can be outlined as, optimization problem preparation, decision variables description, harmony memory (HM) initialization, generation of a new harmony solution and HM updating [81]. The basic elements of HAS are harmony, HM, HM size (HMS), maximum iterations, HM considering ratio (HMCR), pitch considering rate (PCR) and fret width (FW). Here, harmony is defined as a set of values corresponding to an objective function and the place where these values are stored is abbreviated as HM. Generally, the best harmony is stored in the first place and the rest are classified according to their performance. This arrangement helps in depicting the HMS and is considered as an important aspect while calibrating a model. A brief overview of functioning of harmony search optimization is depicted in Figure 8.3.

Further, HS algorithm is implemented for identifying the wavelet coefficients of discrete WT for the purpose of developing an efficient fault classification mechanism.

8.4.3   Statistical features

This section develops a wavelet-based fault classification technique, by choosing an optimum combination of mother wavelets and the number of wavelet decomposition levels. This combination helps in extracting the most important attributes from a given signal, which can be trained for fault classification for PV systems.

8.4.3.1   Feature vector representation

8.4.3.1.1   Signal preprocessing

In signal preprocessing, the voltage and current parameters extracted from various faults in PV systems are considered. Initially, these signals are sampled at a rate of 64 samples per cycle, which is mostly used in digital protective relays as reported by [82]. To remove the steady state information from the signal and keep only the

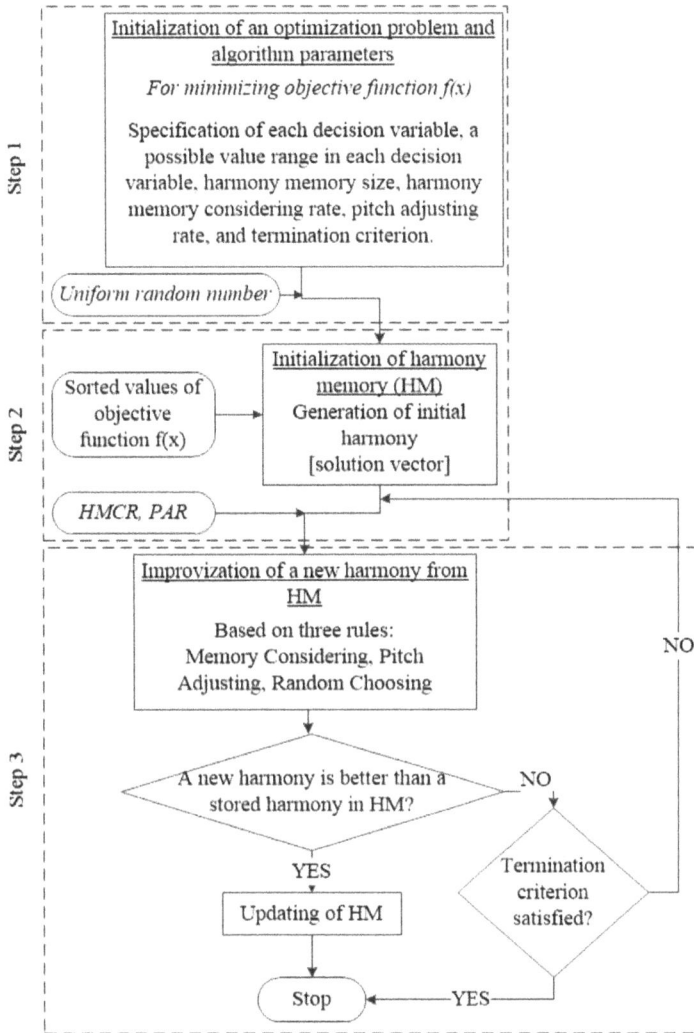

Figure 8.3   HS optimization algorithm

transient information following the fault inception, the difference between the samples in each two successive cycles as shown in Figure 8.4 is calculated using the following:

$$V^D(j) = V(j + n) - V(j) \quad j = 1, 2, \ldots, n \tag{8.7}$$

where V is the type of the signal (current or voltage).

The discrete WT is applied to the sequences ($_xD$) of the currents and voltages. Equations (8.3)–(8.6) are utilized to calculate the approximation and detail wavelet coefficients. The number of decomposition levels is selected to be four to guarantee

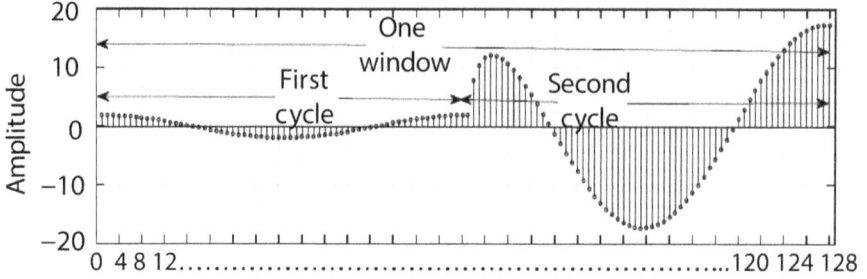

(a)

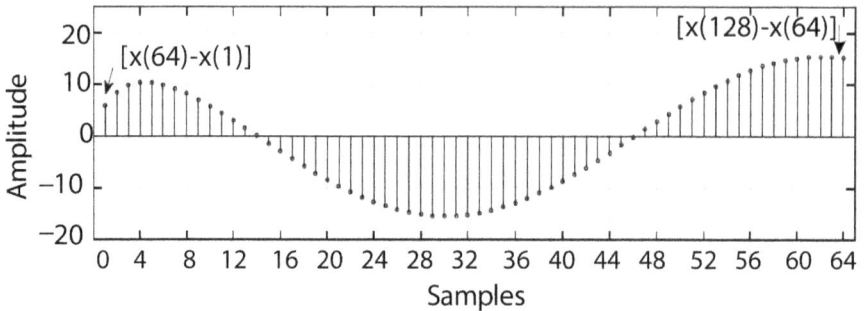

(b)

Figure 8.4   *The analysis window of the sampled signal: (a) one window of two cycles, (b) the difference between two cycles*

that the power system frequency (i.e., 50 Hz) is centred at the approximation level (i.e., Ca_4), which spans between 0 and 100 Hz as shown in Figure 8.5.

The outcome of the WT analysis is a vector consisting of the wavelet coefficients, which is then used to compute the energy of the wavelet coefficients of the details ($g_{d1} - g_{d4}$) and the approximation (g_{d4}) for each current or voltage sampled signals. The steps of implementing the DWT analysis, which are described in Section 8.4.1, are then repeated for various types of faults to calculate the energies of the wavelet coefficients for each fault type and then tabulate them into one array W_c, which consists of the all-fault types coefficients energies.

8.4.3.1.2   Normalization

The energy of the wavelet coefficients, which are described by the vector as depicted in Figure 8.6, usually needs normalization. There is a need for normalization to ensure that, there is no unbalance in the values of the energy of the wavelet coefficients that may arise due to the large values of the approximation coefficients compared to those at the detail levels. To address this issue, the energies vector (E_{ag})

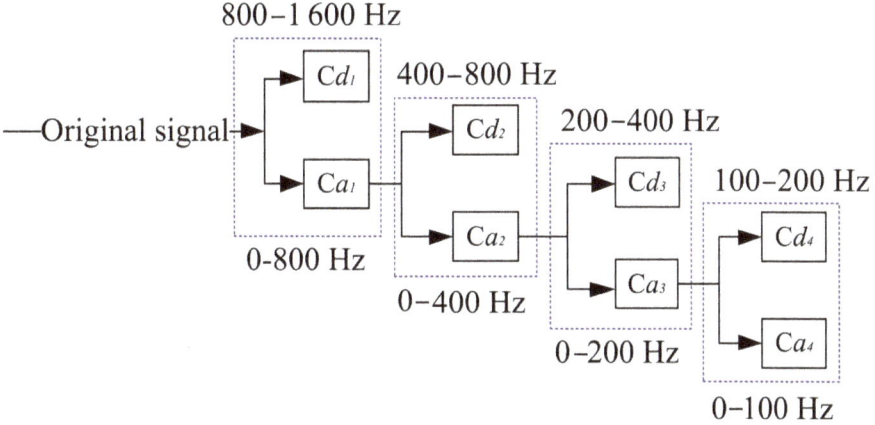

Figure 8.5   Wavelet decomposition levels for signal preprocessing

of the wavelet coefficients is normalized to get the normalized value (ZE_{ag}) of the energy of the wavelet coefficient vector and the remaining elements in the energy's vectors for all other fault types are computed in a similar way.

$$Z_{Eag}(k) = \frac{E_{ag}(k) - \mu(E_{ag})}{\sigma(E_{ag})} \tag{8.8}$$

8.4.3.1.3   Reference vector

Each type of fault is represented by a vector, which holds the energy of the wavelet coefficients. To represent each fault by only one value, another vector was generated, which uses as a reference vector. Consequently, the distances between the vectors for all other fault types and this vector (reference vector) can be easily calculated.

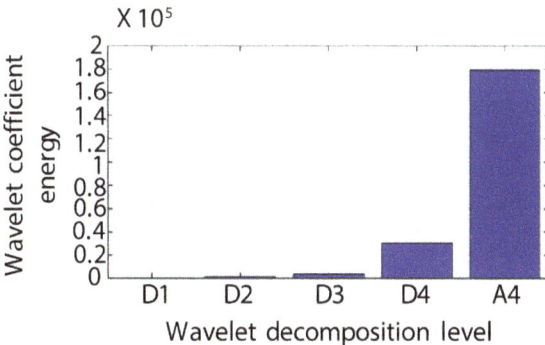

Figure 8.6   Energy vector of wavelet coefficients

To generate a reference vector, the DWT is applied to the current and voltage signals at a non-fault (i.e., healthy) case and then the wavelet coefficients energies of that non-faulty case are sorted in a vector E_h and then normalized to get Z_{Eh} using (8.9).

$$Z_{Eh}(k) = \frac{E_h(k) - \mu(E_h)}{\sigma(E_h)} \tag{8.9}$$

8.4.3.1.4   Euclidean distance
The distance among the vector Z_E of every fault and the non-faulty condition vector Z_{Eh} is computed with (8.10).

$$ED_t \sqrt{\sum r \left[Z_{gt}(r) - Z_{gh}(r) \right]^2} \tag{8.10}$$

All the values of the Euclidean distances among the fault types are sorted in a single vector D_t.

8.4.3.1.5   Variance
The variance $\sigma_{D_t}^2$ between the distances in the distance vector D_t is calculated as in (8.11):

$$\sigma_{D_t}^2 = \frac{1}{m-1} \sum P \left[D_t(P) - \mu(D_t) \right]^2 \tag{8.11}$$

Hence, the value of the variance $\sigma_{D_t}^2$ needs to be maximized for maximum differences between all fault types. Moreover, the variance is affected by the wavelet functions combination choice used in the analysis. To overcome this problem HSA is utilized.

In this work, the variance $\sigma_{D_t}^2$ represents the objective function which is to be maximum.

$$Maximize \ \{\sigma_{D_t}^2\} \tag{8.12}$$

The decision variable set is y, the total number of decision variables is N_y, each decision variable y_x are restricted by the maximum value y_{max} and minimum value y_{min}. The HSA randomly allocates the values for each decision variable. In every iteration, the decision variable vector y consists of various elements, where each element is allocated a numeric value. At the beginning, the HSA algorithm arbitrarily generates the initial values for every decision variable as in (8.13):

$$y_x = y_{xmin} + rand_1 \left(y_{xmax} - y_{xmin} \right) \tag{8.13}$$

where $rand_1$ is an arbitrary number created using the uniform distribution between 0 and 1. Equation (8.14) depicts the decision variables arranged in the HM matrix, where N_y is the number of decision variables and H_s is the HMS, which is an input parameter to the algorithm.

$$
\text{HM} = \begin{bmatrix} Y_1^1 & \cdots & Y_{Ny}^1 \\ \cdots & \cdots & \cdots \\ Y_1^{H_s} & \cdots & Y_1^{H_s} \end{bmatrix} \rightarrow \begin{bmatrix} f_1(y) \\ \cdots \\ f_{Hs}(y) \end{bmatrix} \tag{8.14}
$$

The values of the fitness function are computed using each row in HM and are then stored in a vector $[f_1(y), \ldots, f_{Hs}(y)]^T$. The main objective of the HSA is to search through multiple trials for the optimal values of decision variables by maximizing the value of fitness function. Hence, in each trial the HSA finds the highest value of the fitness function and tries to replace it with another new one with a lower value.

A new solution vector with new values for the decision variable values is generated as $y^{new} = [y_1^{new}, \ldots, y_{Ny}^{new}]$. The solution vector considers HM rate HMC_r, the pitch adjusting rate PA_r and the bandwidth (bw). Each value of the decision variable in the new vector y^{new} is generated according to (8.15).

$$
y_1^{new} \leftarrow \begin{cases} y_1^{new} \in [y_1^1, \ldots, y_1^{Hs}] & rand_2 \leq HMC_r \\ y_1^{new} \in [y_{1max}, y_{1max}] & rand_2 \leq HMC_r \end{cases} \tag{8.15}
$$

where $rand_2$ is an arbitrary number created using the uniform distribution between 0 and 1. The new decision variable value is adjusted according to PA_r.

$$
y_1^{new} \leftarrow \begin{cases} y_1^{new} \pm rand \times bw & rand_3 \leq PA_r \\ y_1^{new} & rand_3 > PA_r \end{cases} \tag{8.16}
$$

where $rand_3$ is an arbitrary number created using the uniform distribution between 0 and 1. The HSA parameters bw and PAr are updated in every epoch to improve the performance of HSA using [83]:

$$
PA_r = PA_{rmin} + \left[(PA_{rmax} - PA_{rmin}) \left(\frac{I_r}{MI_r} \right) \right] \tag{8.17}
$$

$$
bw = bw_{max} \times e^{(\ln [bw_{min}/bw_{max}/MIr] \times Ir)} \tag{8.18}
$$

where PA_{rmin} = minimum pitch adjusting rate, PA_{rmax} = maximum pitch adjusting rate, bw_{min} = minimum bandwidth, bw_{max} = maximum bandwidth, I_r = current iteration and MI_r = maximum iteration.

If the value of objective function for the new harmony vector is less than the worst member value, the HSA change (update) the HM matrix by replacing the worst harmony vector by the new one. The db5 wavelet coefficients obtained through harmony search were applied to the voltage signals of various PV faults and operating conditions.

8.4.3.2 Feature extraction

Once the wavelet coefficients and level of decomposition and reconstruction are determined, the feature extraction process can be initiated. During this process, the most relevant information from different signals is extracted and represented in a lower dimensional space. This relevant information regarding the signals helps in discriminating and distinguishing different objects or faults [84].

Spectral analyser is widely used in the first step of the feature detection process. The spectrogram can be useful in gathering information about a signal and in determining which features can be extracted from a signal. The spectrogram is used for defining statistical features of the signal, such as mean value, standard deviation, skewness and kurtosis. These features build a feature vector to retrieve similar signal from the database. Some of the other features that can be extracted are peak to peak value, power, energy, entropy, total harmonic distortion and signal to noise ratio. The process of feature extraction using wavelets is explained in [21].

Feature extraction process is initiated after the level of decomposition and reconstruction is performed with the wavelet coefficient. During the process of feature extraction critical information from the signals are obtained which aids in classification of different fault states [36]. The process of feature extraction is usually performed using spectral analyser. The spectrogram aids in attaining information regarding signal and identification of feature that can be extracted. Statistical feature such as mean value, standard deviation, skewness and kurtosis are defined using spectrogram. A feature vector is formed by the features for retrieving similar signals from the database. Feature that can be extracted are peak to peak value, power, energy, entropy, total harmonic distortion and signal to noise ratio. These features can be expressed as follows:

Mean of the reconstructed signal is calculated by

$$\text{Mean } (\mu) = \frac{1}{N} \sum_{i=0}^{N-1} V_i \tag{8.19}$$

where N denotes the number of samples and V_i is the sampled reconstructed signal.

Standard deviation of the reconstructed signal is calculated by

$$\text{Standard deviation } (\sigma) = \frac{1}{N-1} \sum_{i-0}^{N-1} (V_i - \mu)^2 \tag{8.20}$$

Skewness and kurtosis of the reconstructed signal is given by

$$\text{Skewness } (S) = \frac{\frac{1}{N} \sum_{i=0}^{N} (V_i - \mu)^3}{\left(\sqrt{\frac{1}{N} \sum_{i=0}^{N} (V_i - \mu)^2}\right)^3} \tag{8.21}$$

$$\text{Kurtosis } (K) = \frac{\frac{1}{N} \sum_{i=0}^{N} (V_i - \mu)^4}{\left(\sqrt{\frac{1}{N} \sum_{i=0}^{N} (V_i - \mu)^2}\right)^4} \tag{8.22}$$

Peak to peak value of the reconstructed signal is given by

$$V_{pp} = 2\sqrt{2}\sigma \tag{8.23}$$

Energy of the decomposed signal is formulated using

$$E = \int_{-\infty}^{\infty} |V(t)|^2 \, dt \tag{8.24}$$

Power of the reconstructed signal is given by

$$P = \lim_{N \to \infty} \frac{1}{2T} \oint_{-T}^{T} |V(t)^2| \, dt \tag{8.25}$$

where $V(t)$ is the reconstructed signal.

Entropy is defined as a major tool in information theory. It is also used to estimate the type of wavelet suitable for decomposing and reconstructing a given signal. The entropy of a given signal is found by

$$H(V) = - \sum_{i=1}^{N} p(V_i) \log_{10} p(Vi) \tag{8.26}$$

where $p(V_i)$ is given by probability of sample of voltage signal.

To estimate the total harmonic distortion of a sinusoidal signal in time domain, we use (8.28):

$$
\begin{aligned}
y(t) \quad &= \alpha_0 + \frac{\alpha_2}{2} + \frac{3\alpha_4}{8} + \left(\alpha_1 + \frac{3\alpha_3}{4} + \frac{10\alpha_5}{16} \right) \\
&\sin\omega_0 t - \left(\frac{\alpha_2}{2} + \frac{\alpha_4}{2} \right) \cos 2\omega_0 t - \left(\frac{\alpha_3}{4} + \frac{5\alpha_5}{16} \right) \\
&\sin 3\omega_0 t + \frac{\alpha_4}{8} \cos 4\omega_0 t + \frac{\alpha_5}{16} \sin 5\omega_0 t
\end{aligned}
\tag{8.27}
$$

where α_i =coefficients of Taylor series.

The signal to noise ratio for a given signal is determined by the ratio of reconstructed signal to original signal:

$$\text{SRN} = \frac{\text{Reconstructed signal}}{\text{Original signal}} \tag{8.28}$$

Once the required features of all the faults and operating conditions of PV systems were extracted, we apply principal component analysis (PCA) to minimize the feature set.

8.4.4 Principle component analysis

PCA is a statistical analysis tool that utilizes multiple dimensions data set for minimization and highlighting the similarity within data set. In case of high dimension data, the similarity identification is very difficult hence for such conditions the PCA analyses the data graphically. After the similarity are identified then the data is reduced with insignificant loss of information. These dimensions reduction can be achieved by transforming the original data set into a series of uncorrelated principal components. Principle algorithm for PCA is shown in Figure 8.7.

By applying PCA, the features are minimized into three uncorrelated variables which were in turn utilized with the ML techniques to obtain the trained data for fault classification.

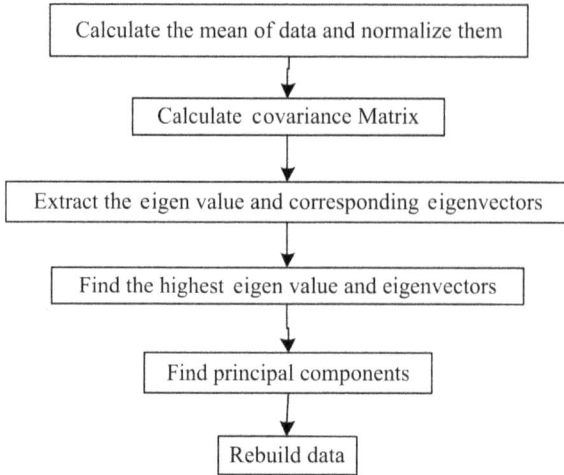

Figure 8.7   Diagram flow of the PCA algorithm

8.5   Machine learning approach

ML is a cluster of algorithms that are used to obtain models from data. There are two main types of ML models. First is the supervised learning model and second is the unsupervised learning model. In this research, the supervised ML model is adapted through K-NNs for the classification process [85].

8.5.1   K-nearest neighbour classifier

The K-NN is a classification learner method which is used for categorizing data into different classes. An assumption in matric space Ω regarding set of points is considered and each point is assigned a label of either 0 or 1. The labelled samples are represented by $(X_1, Y_1), (X_2, Y_2), \ldots (X_n, Y_n)$ and (X, Y) are in query. The label of query is predicted based on the most common class among the k closest point to the labelled sample X as illustrated in the Figure 8.8.

Case of ties has been avoided in the considered example. The figure illustrates two possible cases where ties can occur in algorithm. It is feasible to having multiple classes transpiring in identically frequently among the query of K-NNs. Even equal distance tie from the multiple points in query is possible. Distribution with density has also been discussed in may research for avoiding distance tie. Random selection is also one of the common methods for breaking ties, hence if a voting tie occurs then a random tie is selected from the most common labels and in case of a distance tie, a random point is selected at a distance. To avoid issues of binary classification, the vote tie is broken by selecting the label 1. By generating random variable $U_1, U_2, \ldots, U_n$, distance tie is broken in case of uniform distribution on [0,1]. In case of distance tie between two point X_i and X_j, X_i is selected if $U_i > U_j$ and X_j is selected if $U_j > U_i$. The pseudocode is presented in algorithm as follows.

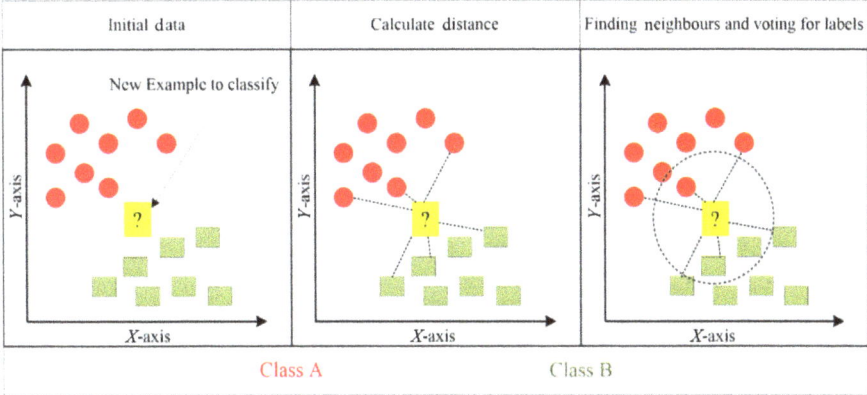

| Initial data | Calculate distance | Finding neighbours and voting for labels |

Class A   Class B

Figure 8.8   Illustration of K-NN

Algorithm: *k*-NN pseudocode

Require: $k \in N$, X is a domain, Y is the response (must be a finite set $\{1,2,\ldots,p\}$),

$\quad \alpha \in X \left(x_1, y_1 \right), \ldots, \left(x_n, y_n \right) \in X \times Y$

{Calculate input point distance from all the data points}
  for $i = 1$ **to** n **do**
  $d_i \leftarrow d(a, x_i)$
  end for
  {Finding *k*-nearest neighbours response for the input point}
  for $i = 1$ **to** k **do**
  $m \leftarrow \arg_m \min \{d_m$ such that $1 \le m \le n$ not previously selected$\}$

  $\quad a_i \leftarrow y_m$

end for
  {Finding the number of iterations}
  for $i = 1$ **to** p **do**
  $d_i \leftarrow d(a, X_i)$
  end for
  $r \leftarrow \{y_i \mid 1 \le i \le p$ such that v_i is maximum among $v_1, v_2, \ldots, v_p\}$
  {identification of most common response among *k* nearest neighbours, and
in case of multiple responses pick a fixed one}
  return r {retuning the common response}

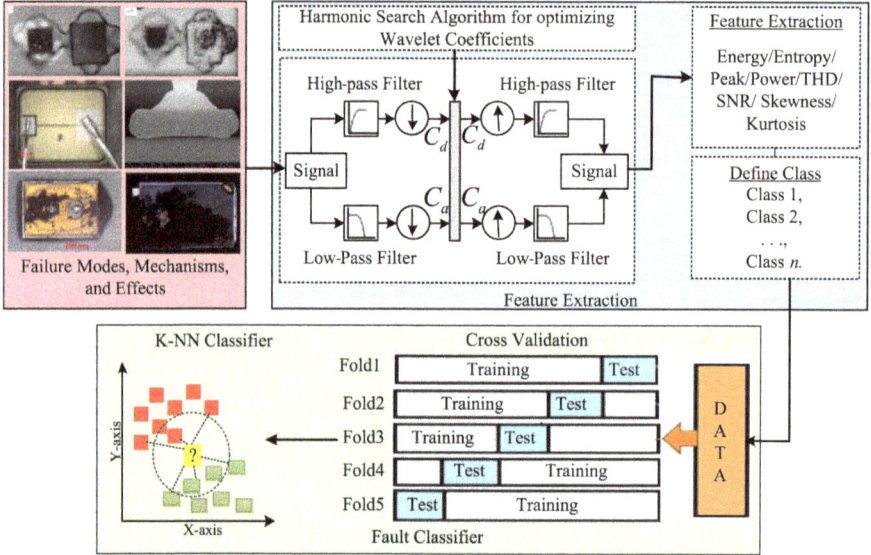

Figure 8.9   Block diagram for the wavelet-based fault detection technique.

8.5.2   Fault classification algorithm

The methodology adapted for fault classification is structured as depicted in Figure 8.9. In this research, the fault classification technique is used to learn a model called a classifier. Data corresponding to various faults and operating conditions of PV system is divided into two groups: training set and testing set. In the training phase, the training set is fed to the classifier for labelled data set into one of the classes depending on the target output. A fivefold cross validation is applied to validate the trained data. In the testing phase, the test samples are verified depending on the target output. Once essential features have been identified, the classification of a fault condition is straightforward performed. The K-NN technique discussed in Section 8.5.1 is used for fault classification due to its advantages with nonlinear data classification as mentioned in Section 8.5.1. For simulation analysis, the input data (i.e., the extracted features) are tabulated and imported to the classification program. The classifier models are trained for functional fault classification.

To observe the performance of the proposed methodology, a 4-kW two-stage PV system for a fixed load is simulated for a varying climatic and load conditions. For experimental purpose, six different modes of operation (normal, bond-wire failure, solder fatigue, overstress due to ESD, wear out due to substrate cracking and other failure conditions) are categorized for the operation of the power PV inverter. The terminal voltage and current measurements of the inverter under different modes of operation for a determined time period are logged. A sample of waveforms recorded for all the operating conditions and failure mechanisms for developing the fault classification algorithm are depicted in Figure 8.10. The process of feature vector representation and applying DWT through HSA for the purpose of feature extraction

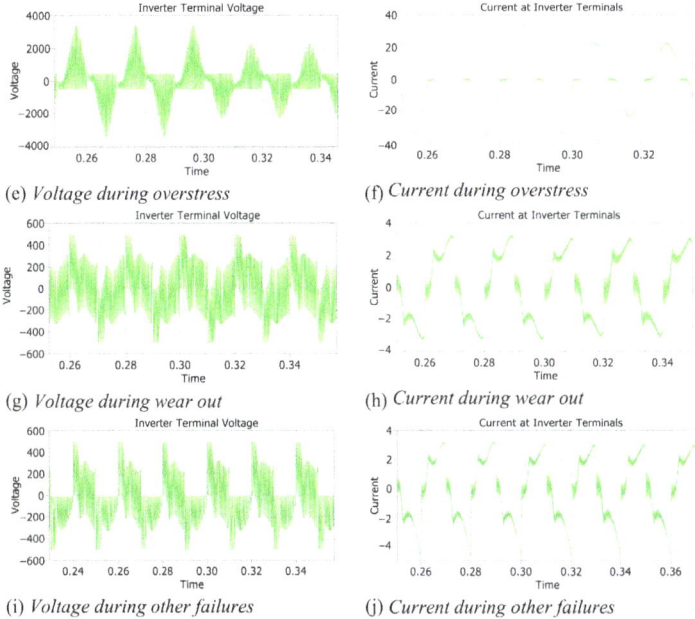

(e) *Voltage during overstress*   (f) *Current during overstress*

(g) *Voltage during wear out*   (h) *Current during wear out*

(i) *Voltage during other failures*   (j) *Current during other failures*

Figure 8.10   *Various operating and fault conditions of PV inverter*

is carried out. Eight different features such as energy, entropy, power, peaks, harmonics, signal to noise ratio, skewness and kurtosis are extracted for voltage and current outputs of each condition. This forms a feature set matrix 4 013 × 8, where the normal operation has 513 × 8 features, bond-wire failure has 1 448 × 8 features, solder fatigue, overstress, wear out and other failures have 513 × 8 features each.

The parameters of the classifier for the training process and the corresponding outcomes are shown in Table 8.5.

The classification accuracy is defined as the ratio of correctly classified samples to the total number of samples in the test data set. The corresponding equation is given as

Table 8.5   *Model type and training performance results for K-NN*

Model type	
Number of neighbours	10
Distance metric	Euclidean
Distance weight	Squared inverse
Standardize data	True
Results	
Accuracy	96.1%
Total misclassification cost	157
Prediction speed	~22 000 obs/s
Training time	4.5238 s

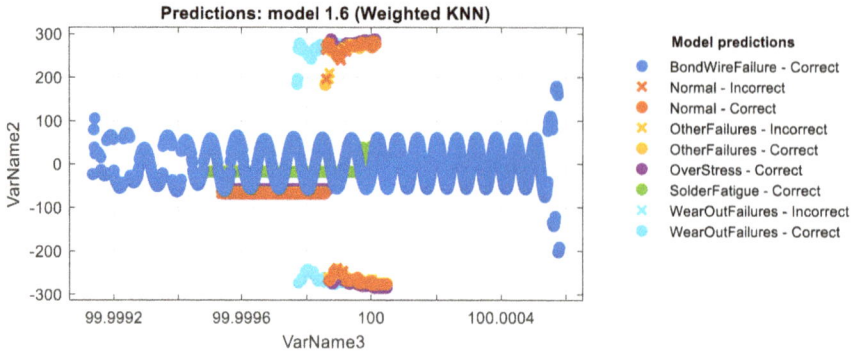

Figure 8.11   Scatter plot of the trained data

$$\text{Accuracy of classification} = \frac{\text{Total no. of samples classified correctly}}{\text{Total no. of samples in the data set}} \times 100 \quad (8.29)$$

To confirm the developed classifier data, a set of input data not used within the training stage, is given to the trained data. This technique improves the average accuracy about 96.1 per cent in training process. The corresponding results were depicted in Figures 8.11–8.13.

Figure 8.11 displays the scatter plot between two features of the training data. The scatter plot, explores the data for important predictors, outliners and visual patterns.

Figure 8.12 depicts the confusion matrix for the experiment. The confusion matrix aids in understanding the performance of the classifier in each of the classes. It supports the understanding that, if the classifier is executed poorly in identifying a class or all the classes are identified accurately. From the result, it is clear that all the classes were classified with almost precision after seeing the confusion matrix. It helps in assessing how currently selected classifier performs in each class.

Figure 8.13 depicts the receiver operating characteristics by plotting the sensitivity (true positive rate) and 1-specificity (false positive rate) for different possible cut points in a training network. It is observed in Figure 8.10 that all the classes are close to the left and top border of the receiver operating characteristic (ROC) space making the training and testing process most accurate.

8.6   Failure mode effect classification analysis

Both FMEA and FMMEA call for a criticality analysis to prioritize failure modes and mechanisms, respectively. Such prioritization allows for efficient allocation of resources for enabling and improving reliability of a system. One difficulty for prioritizing failure mechanisms for component level FMMEA is that the information necessary to make the decision is highly application dependent. This section will describe the traditional method for defining and estimating criticality and establish

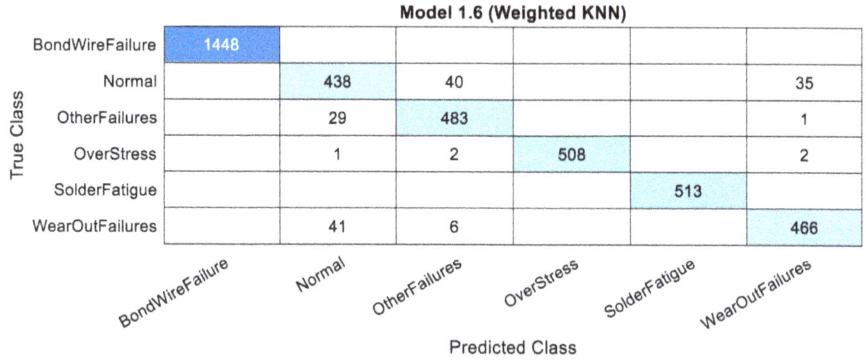

Model 1.6 (Weighted KNN)

True Class	BondWireFailure	Normal	OtherFailures	OverStress	SolderFatigue	WearOutFailures
BondWireFailure	1448					
Normal		438	40			35
OtherFailures		29	483			1
OverStress		1	2	508		2
SolderFatigue					513	
WearOutFailures		41	6			466

Predicted Class

(a) Truly and falsely classified features with weighted K-nearest neighbour

Model 1.6 (Weighted KNN)

True Class	BondWireFailure	Normal	OtherFailures	OverStress	SolderFatigue	WearOutFailures	TPR	FNR
BondWireFailure	100.0%						100.0%	
Normal		85.4%	7.8%			6.8%	85.4%	14.6%
OtherFailures		5.7%	94.2%			0.2%	94.2%	5.8%
OverStress		0.2%	0.4%	99.0%		0.4%	99.0%	1.0%
SolderFatigue					100.0%		100.0%	
WearOutFailures		8.0%	1.2%			90.8%	90.8%	9.2%

Predicted Class

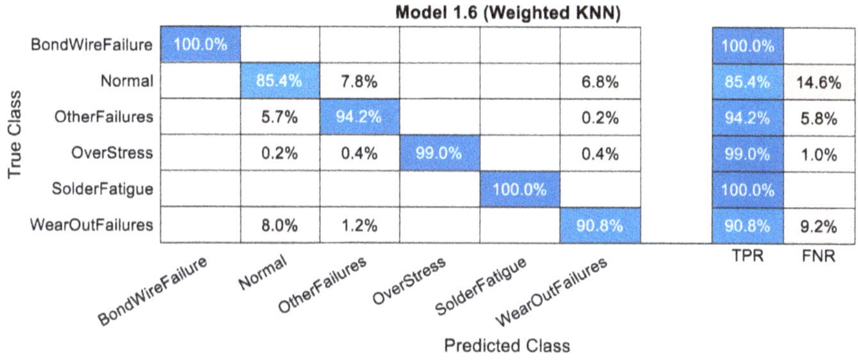

(b) True positive rate (TPR) and false negative rate (FNR) of features with weighted K-nearest neighbour

Model 1.6 (Weighted KNN)

True Class	BondWireFailure	Normal	OtherFailures	OverStress	SolderFatigue	WearOutFailures
BondWireFailure	100.0%					
Normal		86.1%	7.5%			6.9%
OtherFailures		5.7%	91.0%			0.2%
OverStress		0.2%	0.4%	100.0%		0.4%
SolderFatigue					100.0%	
WearOutFailures		8.1%	1.1%			92.5%
PPV	100.0%	86.1%	91.0%	100.0%	100.0%	92.5%
FDR		13.9%	9.0%			7.5%

Predicted Class

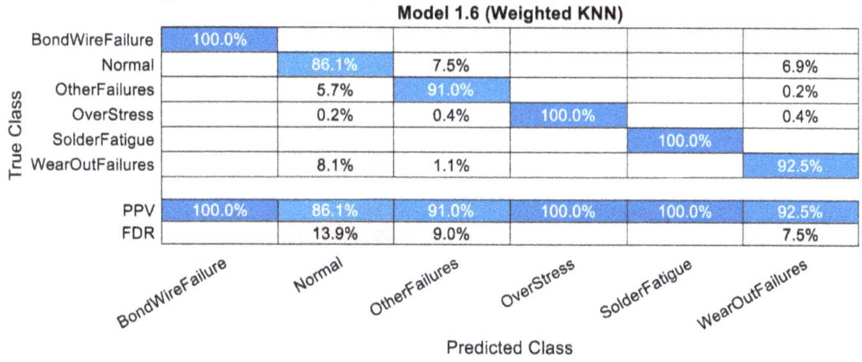

(c) Positive predictive values (PPV) and false discovery rate (FDR) of classification process

Figure 8.12   Confusion matrix for trained data

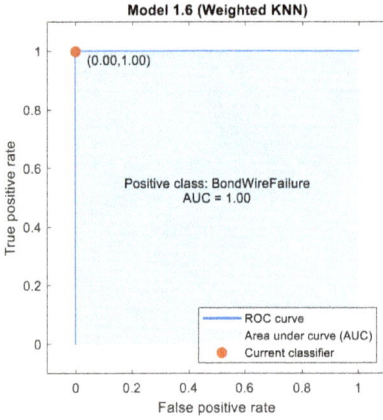

(a) ROC-AUC for bond-wire failure

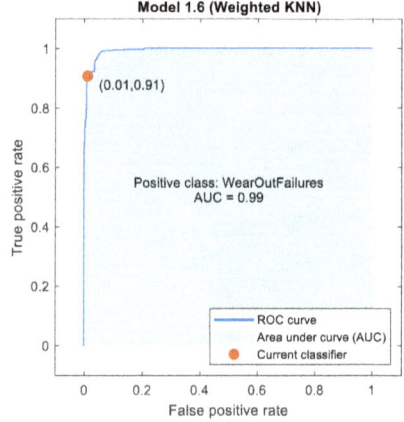

(b) ROC-AUC for wear out failure

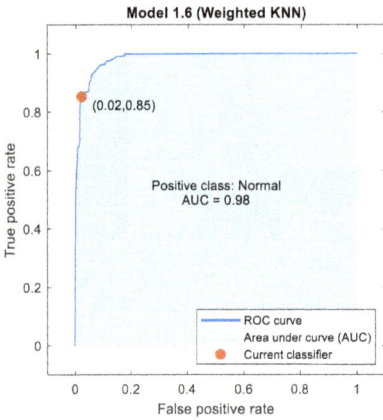

(c) ROC-AUC for normal operation

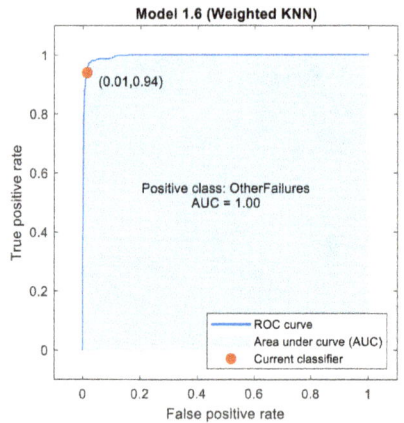

(d) ROC-AUC for other failures

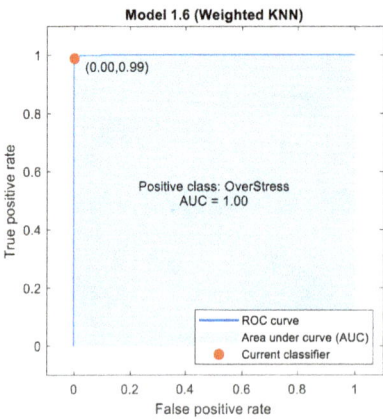

(e) ROC-AUC for over stress failure

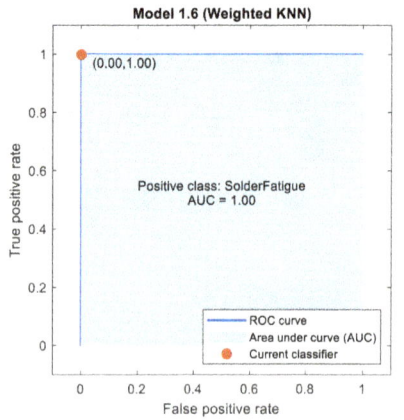

(f) ROC-AUC for solder fatigue

Figure 8.13   ROC for trained data

component-level information-based guidance for ranking failure mechanisms based on criticality.

8.6.1 Approaches for criticality analysis

Through JEP131B, JEDEC outlines three components for critically ranking failure modes: severity, occurrence and detection [48]. Each of the three categories is separately given a ranking from 1 to 10 based on the judgement of the team that is completing the analysis with 10 being the most severe, highest occurrence and most difficult to detect. These three metrics are then multiplied together create a risk priority number (RPN). Failure modes with higher RPNs are determined to be of more concern than those with lower RPNs. Corrective actions meant to be prioritized to lower the RPN of the highest failure modes. The RPN should be updated after corrective actions are taken and the design should be re-evaluated. As the rankings are based on the judgement of the team, RPN rankings should not be compared with another group's RPN ranking for the same or any other system.

The severity of a failure mode is dependent on the effects to the end user. First and foremost, severity should consider any potential to harm the users of a system. If there is potential to harm due to the effects from a certain mode (or mechanism), that mode (or mechanism) should be assigned a higher severity rating. The next considerations should include costs to the user and the system manufacturer. Costs take on a variety of forms but may include legal, warranty and returns, associated maintenance and brand reputation. Based on the judgement of the team a severity ranking should be given which takes into consideration these factors. It is evident that the severity of the failure is application dependent. Component level severity will be discussed subsequently.

The occurrence of a failure mode is how likely it is to occur. Considerations for occurrence should include environmental and loading conditions as well as system materials, geometries and part types. Using this information, it is possible to establish a probability of a mode or mechanism that can then be ranked according to the judgement of the FMEA development team. This information is also application dependent.

Finally, the detection metric for a failure mode is traditionally defined as the ability to detect a failure mode before shipping the product to the customer. Traditionally, in the electronics industry detection deals with the escape rate of any given test or screen. JEDEC suggests using inverse of the escape rate as one way of quantifying the detection of the mode. As the scope of the FMEA developed for a system level, individual power semiconductors are not assumed to be tested at the system integration level. Therefore, a different approach to detection will be taken in the subsequent discussion.

8.6.2 Approaches for severity analysis

Severity is determined by the effect to the end user. In the absence of this information, severity must be viewed in a different context. However, all Si power devices will be used in a larger circuit and thus have other electrical components nearby. For the purposes of a component level prioritization, severity will be determined by the potential of the failure to be catastrophic and effect nearby components.

With respect to Si power devices, overstress mechanisms which result in short circuit failure have the most potential to damage the system around them. Short circuits can create significant joule heating, which, if uncontrolled can damage nearby components and potentially start a fire thus increasing the associated costs of failure. The wear out mechanisms can be considered less severe because it is likely that only the Si power device fails, not harming other nearby components, and the system can potentially be repaired or replaced.

8.6.3　Occurrence

Occurrence is dependent on the application operating and environmental conditions. Certain failure mechanisms can only be expected to occur when specific stressors are present. For example, electrochemical migration is not a concern in an application where humidity is below the threshold for initiation. To calculate a value for an occurrence there are two approaches: one for wear out mechanisms and one for overstress mechanisms. For wear out mechanisms, one must identify a failure model which relates the stressors with the materials and geometries of the system, from this a time to failure or equivalent can occur giving an indication of the occurrence of the mechanism in the application. One example of this is the use of the Norris–Landsberg model for calculating fatigue of die attach. Failure models express time-to-failure, or equivalent, as a function of the stresses action on a system. Overstress failures are given a high priority with respect to occurrence as the stresses which are reasonably expected in the life cycle profile should be designed against. Assuming the proper design precautions have been taken, overstress mechanisms are unlikely to occur in the field and can only be quantified by identifying a probability that an overstress condition could occur. For example, the probability that a lightning strike causes a burnout of a device due to overcurrent can be a 'measure' of occurrence.

8.6.4　Detection

Detection in the traditional sense is determined by the ability to detect a failures, defects and non-conformities before it leaves manufacturing or assembly. For a component level discussion, this traditional definition is not applicable. Therefore, detection will be considered as the ability to detect a failure in operation, before the failure occurs. For an expected loading profile, overstress mechanisms can be avoided through proper selection of parts and appropriate de-rating. However, overstress failures may still occur due to random and unpredictable loading excursions such as lightning strikes or crashes.

Not all mechanisms are unpredictable as accumulating damage changes some observable and measurable parameters. The ability to monitor and predict failure is a field of study referred to as prognostics and health management (PHM). In PHM, in situ data is monitored and analysed for purposes of anomaly detection, fault classification and remaining useful life calculation. Wear out mechanism can be detectable and several groups have successfully implemented PHM for Si power devices [50,

86, 87]. Certain wear out mechanisms are more detectable than other wear out and overstress mechanisms depending on the feasibility and correlation with damage of electrical parameters associated with the mechanism. PHM techniques and methods are developing rapidly and reducing in cost and ease of implementation; however, the development and implementation of a PHM framework is not yet trivial and therefore it may only be cost efficient for certain critical components and applications. Additionally, competing failure mechanisms may convolute the measurement signals as they have the same or similar failure modes making it difficult to distinguish between the failure modes and take necessary corrective action.

8.7 Summary

This chapter developed a failure mode classification mechanism for condition monitoring of PV inverters. The developed algorithm performed signal preprocessing by DWT for noise removal, feature extraction and region of interest segmentation. The wavelet coefficients associated with the DWT were optimized by a novel approach based on HSA. Various types of features were extracted once the signal preprocessing is completed. The harmony search analysis proved to be very efficient in choosing the best wavelet coefficient depending upon the structure of the signal. The extracted features are assigned towards corresponding classes and randomly divided as training and test data for the purpose of evaluation of the classifier. K-NN is used to classify the fault conditions of PV inverters into normal and faulty status. A fivefold cross validation is performed to measure the performance of the classifier with input data. On validation, the developed approach depicted a training accuracy of 96.1 per cent. Further, criticality analysis is established from component-level information-based guidance for ranking failure mechanisms.

References

[1]   Koutroulis E., Blaabjerg F. 'Design optimization of transformerless grid-connected PV inverters including reliability'. *IEEE Transactions on Power Electronics*. 2013;**28**(1):325–35.

[2]   Yang Y., Sangwongwanich A., Blaabjerg F. 'Design for reliability of power electronics for grid-connected photovoltaic systems'. *CPSS Transactions on Power Electronics and Applications*. 2016;**1**(1):92–103.

[3]   Shahzad M., Bharath K.V.S., Khan M.A., Haque A. 'Review on reliability of power electronic components in photovoltaic inverters'. 2019 International Conference on Power Electronics, Control and Automation (ICPECA); 2019. pp. 1–6.

[4]   Wang H., Zhou D., Blaabjerg F. 'A reliability-oriented design method for power electronic converters'. 2013 Twenty-Eighth Annual IEEE Applied Power Electronics Conference and Exposition; 2013. pp. 2921–8.

[5]   Bahman A.S., Iannuzzo F., Blaabjerg F. 'Mission-profile-based stress analysis of bond-wires in sic power modules'. *Microelectronics Reliability*. 2016;**64**(4):419–24.

[6]   Kurukuru V.S.B., Haque A., Khan M.A., Tripathy A.K. 'Reliability analysis of silicon carbide power modules in voltage source converters'. 2019 International Conference on Power Electronics, Control and Automation; 2019. pp. 1–6.

[7]   Lee K.-W., Kim M., Yoon J., Lee S.B., Yoo J.-Y. 'Condition monitoring of DC-link electrolytic capacitors in adjustable-speed drives'. *IEEE Transactions on Industry Applications*. 2008;**44**(5):1606–13.

[8]   Haque A. 'Maximum power point tracking (MPPT) scheme for solar photovoltaic system'. *Energy Technology & Policy*. 2014;**1**(1):115–22.

[9]   Ristow A., Begovic M., Pregelj A., Rohatgi A. 'Development of a methodology for improving photovoltaic inverter reliability'. *IEEE Transactions on Industrial Electronics*. 2008;**55**(7):2581–92.

[10]   Haque A., Bharath K.V.S., Khan M.A., Khan I., Jaffery Z.A. 'Fault diagnosis of photovoltaic modules'. *Energy Science Engineering*. 2019;**3**:255.

[11]   Kurukuru V.S.B., Blaabjerg F., Khan M.A., Haque A. 'A novel fault classification approach for photovoltaic systems'. *Energies*. 2020;**13**(2):308.

[12]   Mahmud N., Zahedi A., Mahmud A. 'A cooperative operation of novel PV inverter control scheme and storage energy management system based on ANFIS for voltage regulation of grid-tied PV system'. *IEEE Transactions on Industrial Informatics*. 2017;**13**(5):2657–68.

[13]   Zhao Y., Lehman B., De Palma J.F., Mosesian J., Lyons R. 'Challenges to overcurrent protection devices under LINE-LINE faults in solar photovoltaic arrays'. 2011 IEEE Energy Conversion Congress and Exposition; 2011. pp. 20–7.

[14]   Pradeep Kumar V.V.S., Fernandes B.G. 'A fault-tolerant single-phase grid-connected inverter topology with enhanced reliability for solar PV applications'. *IEEE Journal of Emerging and Selected Topics in Power Electronics*. 2017;**5**(3):1254–62.

[15]   Mendes A.M.S., Marques Cardoso A.J. 'Voltage source inverter fault diagnosis in variable speed AC drives, by the average current Park's vector approach'. *IEEE International Electric Machines and Drives Conference. IEMDC'99. Proceedings (Cat. No.99EX272)*:704–6.

[16]   Gilreath P., Singh B.N. 'A new centroid based fault detection method for 3-phase inverter-fed induction motors'. IEEE 36th Conference on Power Electronics Specialists; 2005. pp. 2664–9.

[17]   Peuget R., Courtine S., Rognon J.-P. 'Fault detection and isolation on a PWM inverter by knowledge-based model'. *IEEE Transactions on Industry Applications*. 1998;**34**(6):1318–26.

[18]   Singh R., Bhushan B. 'Improving self-balancing and position tracking control for ball balancer application with discrete wavelet transform-based fuzzy logic controller'. *International Journal of Fuzzy Systems*. 2021;**23**(1):27–41.

[19]   Singh R., Bhushan B. 'Improved ant colony optimization for achieving self-balancing and position control for balancer systems'. *Journal of ambient intelligence and humanized computing*. 2020;**53**(2020):1–18.

[20]   Singh R., Bhushan B. 'Condition monitoring based control using wavelets and machine learning for unmanned surface vehicles'. *IEEE Transactions on Industrial Electronics*. 2020;**68**(8):1):7464:73.

[21]   Ahmad S., Hasan N., Bharath Kurukuru V.S., Ali Khan M., Haque A. 'Fault classification for single phase photovoltaic systems using machine learning techniques'. 2018 8th IEEE India International Conference on Power Electronics; 2018. pp. 1–6.

[22]   Kurukuru V.S.B., Haque A., Khan M.A., Tripathy A.K. 'Fault classification with robust knowledge transfer for single phase grid connected photovoltaic systems'. 2019 International Conference on Power Electronics, Control and Automation; 2019. pp. 1–6.

[23]   Khan M.A., Kurukuru V.S.B., Haque A., Mekhilef S. 'Islanding classification mechanism for grid-connected photovoltaic systems'. *IEEE journal of emerging and selected topics in power electronics*. 2021;**9**(2):1966–75.

[24]   Heidari M., Seifossadat G., Razaz M. 'Application of decision tree and discrete wavelet transform for an optimized intelligent-based islanding detection method in distributed systems with distributed generations'. *Renewable and Sustainable Energy Reviews*. 2013;**27**(4):525–32.

[25]   Hopwood M., Gunda T., Seigneur H., Walters J. 'An assessment of the value of principal component analysis for photovoltaic IV trace classification of physically-induced failures'. 2020 47th IEEE Photovoltaic Specialists Conference; 2020. pp. 0798–802.

[26]   Khoa N.M., Viet D.T., Hieu N.H. 'Classification of power quality disturbances using wavelet transform and k-nearest neighbor classifier'. 2013 IEEE International Symposium on Industrial Electronics; 2013. pp. 1–4.

[27]   Khan M.A., Haque A., Kurukuru V.S.B. 'An efficient islanding classification technique for single phase grid connected photovoltaic system'. 2019 International Conference on Computer and Information Sciences; 2019. pp. 1–6.

[28]   Pandarakone S.E., Masuko M., Mizuno Y., Nakamura H. 'Deep neural network based bearing fault diagnosis of induction motor using fast Fourier transform analysis'. 2018 IEEE Energy Conversion Congress and Exposition; 2018. pp. 3214–21.

[29]   Vernica I., Wang H., Blaabjerg F. 'Uncertainties in the lifetime prediction of IGBTs for a motor drive application'. 2018 IEEE International Power Electronics and Application Conference and Exposition; 2018. pp. 1–6.

[30]   Wu R., Blaabjerg F., Wang H., Liserre M., Iannuzzo F. 'Catastrophic failure and fault-tolerant design of IGBT power electronic converters: an overview'. IECON 2013 – 39th Annual Conference of the IEEE Industrial Electronics Society; 2013. pp. 507–13.

[31]   Gao Z., Cecati C., Ding S.X. 'A survey of fault diagnosis and fault-tolerant techniques – Part 1&2: Fault diagnosis with knowledge-based and hybrid/active approaches'. *IEEE Transactions on Industrial Electronics: A Publication of the IEEE Industrial Electronics Society*. 2015;**62**(6):3768–74.

[32] Rothenhagen K., Fuchs F.W. 'Performance of diagnosis methods for IGBT open circuit faults in voltage source active rectifiers'. *IEEE 35th Annual Power Electronics Specialists Conference (IEEE Cat. No.04CH37551)*; 2004. pp. 4348–54.

[33] Yang S., Xiang D., Bryant A., Mawby P., Ran L., Tavner P. 'Condition monitoring for device reliability in power electronic converters: a review'. *IEEE Transactions on Power Electronics*. 2010;**25**(11):2734–52.

[34] Vidhya S D., M B, Balaji M. 'Failure-mode analysis of modular multilevel capacitor-clamped converter for electric vehicle application'. *IET Power Electronics*. 2019;**12**(13):3411–21.

[35] Choi U.-M., Blaabjerg F. 'Separation of wear-out failure modes of IGBT modules in grid-connected inverter systems'. *IEEE Transactions on Power Electronics*. 2018;**33**(7):6217–23.

[36] DiazR.P., Wang H., Yang Y., Blaabjerg F. 'Prediction of bond wire fatigue of IGBTs in a pv inverter under a long-term operation'. *IEEE transactions on power electronics*. 2015;**31**(10):1):7171:82.

[37] Gorecki K., Gorecki P., Zarebski J. 'Measurements of parameters of the thermal model of the IGBT module'. *IEEE Transactions on Instrumentation and Measurement*. 2019;**68**(12):4864–75.

[38] Khatir Z., Lefebvre S. 'Thermal analysis of power cycling effects on high power IGBT modules by the boundary element method'. Seventeenth Annual IEEE Semiconductor Thermal Measurement and Management Symposium (Cat. No.01CH37189); 2001. pp. 27–34.

[39] Anurag A., Yang Y., Blaabjerg F. 'Thermal performance and reliability analysis of single-phase PV inverters with reactive power injection outside feed-in operating hours'. *IEEE Journal of Emerging and Selected Topics in Power Electronics*. 2015;**3**(4):870–80.

[40] Khan M.A., Mishra S., Haque A. 'A present and future state-of-the-art development for energy-efficient buildings using PV systems'. *Intell. Build. Int.* 2018:1–20.

[41] Stevanovic L.D., Beaupre R.A., Gowda A.V., Pautsch A.G., Solovitz S.A. 'Integral micro-channel liquid cooling for power electronics'. 2010 Twenty-Fifth Annual IEEE Applied Power Electronics Conference and Exposition (APEC); 2010. pp. 1591–7.

[42] Wang P., McCluskey P., Bar-Cohen A. 'Hybrid solid- and liquid-cooling solution for isothermalization of insulated gate bipolar transistor power electronic devices'. *IEEE Transactions on Components, Packaging and Manufacturing Technology*. 2013;**3**(4):601–11.

[43] Mohamed S., Jeyanthy P., Devaraj D., Shwehdi M., Aldalbahi A. 'DC-Link voltage control of a grid-connected solar photovoltaic system for fault ride-through capability enhancement'. *Applied Sciences*. 2019;**9**(5):952.

[44] Zhu B., Tan C., Farshadnia M., Fletcher J.E. 'Postfault zero-sequence current injection for open-circuit diode/switch failure in open-end winding PMSM machines'. *IEEE Transactions on Industrial Electronics*. 2019;**66**(7):5124–32.

[45] Department of Defense. *MIL-STD-1629A (Procedures for performing a Failure Mode, Effects and Criticality Analysis*; 1940.

[46] Society of Automotive Engineers. *Fault/Failure Analysis Procedure ARP926C*; 2018.

[47] JESD88E. *JEDEC Dictionary of Terms for Solid-State Technology*; 2013.

[48] JEDEC Solid State Technology Association. JEP131A-*Potential Failure Mode and Effects Analysis (FMEA)*, Arlington. 2005. Available from http://www.jedec.org/Catalog/catalog.cfm.

[49] Mathew S., Alam M., Pecht M. 'Identification of failure mechanisms to enhance prognostic outcomes'. *Journal of Failure Analysis and Prevention*. 2012;**12**(1):66–73.

[50] Patil N., Das D., Yin C., Lu H., Bailey C., Pecht M. 'A fusion approach to IGBT power module prognostics'. EuroSimE 2009 – 10th International Conference on Thermal, Mechanical and Multi-Physics Simulation and Experiments in Microelectronics and Microsystems; 2009. pp. 1–5.

[51] Smet V., Forest F., Huselstein J.-J., *et al.* 'Ageing and failure modes of IGBT modules in high-temperature power cycling'. *IEEE Transactions on Industrial Electronics*. 2011;**58**(10):4931–41.

[52] Lefebvre S., Joubert P.-Y., Labrousse D., Bontemps S. 'Estimating current distributions in power semiconductor dies under aging conditions: bond wire liftoff and aluminum reconstruction'. *IEEE Transactions on Components, Packaging and Manufacturing Technology*. 2015;**5**(4):483–95.

[53] Martineau D., Mazeaud T., Legros M., Dupuy P., Levade C., Vanderschaeve G. 'Characterization of ageing failures on power MOSFET devices by electron and ion microscopies'. *Microelectronics Reliability*. 2009;**49**(9):1330–3.

[54] Navarro L.A., Perpiñà X., Vellvehi M., Banu V., Jordà X. 'Thermal cycling analysis of high temperature die-attach materials'. *Microelectronics Reliability*. 2012;**52**(9–10):2314–20.

[55] Lai W., Chen M., Ran L., Alatise O., Xu S., Mawby P. 'Low stress cycle effect in IGBT power module die-attach lifetime modeling'. *IEEE Transactions on Power Electronics*. 2016;**31**(9):6575–85.

[56] McCluskey P. 'Reliability of power electronics under thermal loading'. 2012 7th International Conference on Integrated Power Electronics Systems (CIPS); 2012. pp. 2011–8.

[57] Frear D. 'Packaging materials' in Kasap S., Capper P. (eds.). *Springer handbook of electronic and photonic materials*. 2017Springer Handbooks. **1**. Springer, Cham: Springer; 2009. pp. 1–17978-3-319-48933-9. Available from https://link.springer.com/chapter/10.1007/978-3-319-48933-9_53.

[58] Fleischer A.S., Chang L., Johnson B.C. 'The effect of die attach voiding on the thermal resistance of CHIP level packages'. *Microelectronics Reliability*. 2006;**46**(5–6):794–804.

[59] Katsis D.C., van Wyk J.D. 'Void-induced thermal impedance in power semiconductor modules: some transient temperature effects'. *IEEE Transactions on Industry Applications*. 2003;**39**(5):1239–46.

[60]   Zhu N. 'Thermal impact of solder voids in the electronic packaging of power devices'. Fifteenth Annual IEEE Semiconductor Thermal Measurement and Management Symposium; 1999. pp. 22–9.

[61]   Ciappa M. 'Selected failure mechanisms of modern power modules'. *Microelectronics Reliability*. 2002;**42**(4–5):653–67.

[62]   Huang J., Hu Z., Gao C., Cui C. Analysis of water vapor control and passive layer process effecting on transistor performance and aluminum corrosion. Proc. 2014 Progn. Syst. Heal. Manag. Conf. PHM 2014; Zhangjiajie, China, 24-27 Aug. 2014; 2014. pp. 26–30 pp..

[63]   Benbahouche L., Merabet A., Zegadi A. 'A comprehensive analysis of failure mechanisms: Latch up and second breakdown in IGBT(IXYS) and improvement'. 2012 19th International Conference on Microwaves, Radar & Wireless Communications; 2012. pp. 190–2.

[64]   Jahdi S., Alatise O., Bonyadi R., *et al.* 'An analysis of the switching performance and robustness of power MOSFETs body diodes: a technology evaluation'. *IEEE Transactions on Power Electronics*. 2015;**30**(5):2383–94.

[65]   Spirito P., Maresca L., Riccio M., Breglio G., Irace A., Napoli E. 'Effect of the collector design on the IGBT avalanche ruggedness: a comparative analysis between punch-through and field-stop devices'. *IEEE Transactions on Electron Devices*. 2015;**62**(8):2535–41.

[66]   Sato M., Kumada A., Hidaka K., Yamashiro K., Hayase Y., Takano T. 'Surface discharges in silicone gel on AlN substrate'. *IEEE Transactions on Dielectrics and Electrical Insulation*. 2016;**23**(1):494–500.

[67]   Fabian J.-H., Hartmann S., Hamidi A. 'Analysis of insulation failure modes in high power IGBT modules'. Fourtieth IAS Annual Meeting. Conference Record of the 2005 Industry Applications Conference; 2005. pp. 799–805.

[68]   Zhang S., Kang R., Pecht M.G. 'Corrosion of ImAg-finished PCBs subjected to elemental sulfur environments'. *IEEE Transactions on Device and Materials Reliability*. 2011;**11**(3):391–400.

[69]   Lu G.-Q., Yang W., Chen X., Chen G., Mei Y.-H.. ' 'Migration of sintered nanosilver on alumina and aluminum nitride substrates at high temperatures in dry air for electronic packaging''. *IEEE Transactions on Device and Materials Reliability: A Publication of the IEEE Electron Devices Society and the IEEE Reliability Society*. 2014;**14**(2):600–6.

[70]   Duvvury C., Rodriguez J., Jones C., Smayling M. 'Device integration for ESD robustness of high voltage power MOSFETs'. Proceedings of 1994 IEEE International Electron Devices Meeting; 1994. pp. 407–10.

[71]   Anolick E.S., Nelson G.R. 'Low-field time-dependent dielectric integrity'. *IEEE Transactions on Reliability*. 1980;**R-29**(3):217–21.

[72]   Su P., Goto K., Sugii T., Hu C. 'A thermal activation view of low voltage impact ionization in MOSFETs'. *IEEE Electron Device Letters: A Publication of the IEEE Electron Devices Society*. 2002;**23**(9):550–2.

[73]   Yan R., Gao R.X. 'Hilbert–Huang transform-based vibration signal analysis for machine health monitoring'. *IEEE Transactions on Instrumentation and Measurement*. 2006;**55**(6):2320–9.

[74] [Hui L., Yuping Z. Bearing faults diagnosis based on EMD and Wigner-Ville distribution. *Proceedings of World Congress on Intelligent Control and Automation*; Dalian, China, 21-23 June 2006; 2006. pp. 5447–51.

[75] Morsi W.G., El-Hawary M.E. 'A new perspective for the IEEE standard 1459-2000 via stationary wavelet transform in the presence of nonstationary power quality disturbance'. *IEEE Transactions on Power Delivery*. 2008;**23**(4):2356–65.

[76] Titchmarsh E. *Introduction to Fourier Integrals*. 2nd Edition. United Kingdom: Oxford University Press; 1948. pp. 1–400.

[77] Mallat S.G. 'A theory for multiresolution signal decomposition: the wavelet representation'. *IEEE Transactions on Pattern Analysis and Machine Intelligence*. 1989;**11**(7):674–93.

[78] Apolloni B., Howlett R.J., Jain L. (eds.). *Knowledge-Based Intell. Inf. Eng. Syst.* 1 Harmony search algorithm for solving Sudoku. **4692**. Berlin, Heidelberg: Springer; 2007.

[79] Parizad A., Khazali A., Kalantar M. 'Application of HSA and GA in optimal placement of facts devices considering voltage stability and losses'. *Electric and Power Energy Conversion System*. 2009;**3**(8):648–54.

[80] Lee K.S., Geem Z.W. 'A new meta-heuristic algorithm for continuous engineering optimization: harmony search theory and practice'. *Computer Methods in Applied Mechanics and Engineering*. 2005;**194**(36–38):3902–33.

[81] Mahdavi M., Fesanghary M., Damangir E. 'An improved harmony search algorithm for solving optimization problems'. *Applied Mathematics and Computation*. 2007;**188**(2):1567–79.

[82] Kezunovic M., Ren J., Lotfifard S. 'Basics of Protective Relaying and Design Principles'. *Design, Modeling and Evaluation of Protective Relays for Power Systems*. 1. Springer, Cham: Springer International Publishing Switzerland 2016; 2016. pp. 1–297.

[83] Livani H., Evrenosoglu C.Y. 'A fault classification method in power systems using DWT and SVM classifier'. *Pes Transmission and Distribution*. 2012:1–5.

[84] Khan M.A., Haque A., Kurukuru V.S.B., Saad M. 'Advanced control strategy with voltage sag classification for single-phase grid-connected photovoltaic system'. *IEEE Journal of Emerging and Selected Topics in Industrial Electronics*. 2020:1–11.

[85] Geethanjali M. Combined wavelet transfoms and neural network (WNN) based fault detection and classification in transmission lines. 2009 International Conference on Control, Automation, Communication and Energy Conservation; 4-6 June 2009, Perundurai, India; 2009. pp. 1–7.

[86] Oh H., Han B., McCluskey P., Han C., Youn B.D. 'Physics-of-failure, condition monitoring, and prognostics of insulated gate bipolar transistor modules: a review'. *IEEE Transactions on Power Electronics*. 2015;**30**(5):2413–26.

[87] Ji B., Song X., Cao W., *et al.* '*In Situ* diagnostics and prognostics of solder fatigue in IGBT modules for electric vehicle drives'. *IEEE Transactions on Power Electronics*. 2015;**30**(3):1535–43.

Index